Safety Instrumented Systems: Practical Probabilistic Calculations

Safety Instrumented Systems Verification: Practical Probabilistic Calculations

William M. Goble
Harry Cheddie

ISA–The Instrumentation, Systems,
and Automation Society

Notice

The information presented in this publication is for the general education of the reader. Because neither the author nor the publisher have any control over the use of the information by the reader, both the author and the publisher disclaim any and all liability of any kind arising out of such use. The reader is expected to exercise sound professional judgment in using any of the information presented in a particular application.

Additionally, neither the author nor the publisher have investigated or considered the affect of any patents on the ability of the reader to use any of the information in a particular application. The reader is responsible for reviewing any possible patents that may affect any particular use of the information presented.

Any references to commercial products in the work are cited as examples only. Neither the author nor the publisher endorse any referenced commercial product. Any trademarks or tradenames referenced belong to the respective owner of the mark or name. Neither the author nor the publisher make any representation regarding the availability of any referenced commercial product at any time. The manufacturer's instructions on use of any commercial product must be followed at all times, even if in conflict with the information in this publication.

Copyright © 2005 ISA - The Instrumentation, Systems and Automation Society

All rights reserved.

Printed in the United States of America.
10 9 8 7 6 5 4 3 2

ISBN 1-55617-909-X

No part of this work may be reproduced, stored in a retrieval system, or transmitted in any form or by any means, electronic, mechanical, photocopying, recording or otherwise, without the prior written permission of the publisher.

ISA
67 Alexander Drive
P.O. Box 12277
Research Triangle Park, NC 27709

Library of Congress Cataloging-in-Publication Data in Process

DEDICATION

*To my wife Janki, son Neil, and daughters Stephanie, Michelle, and Jennifer
for their support, patience and encouragement*
~ Harry L. Cheddie

To those who provided great support, training and mentoring:
Sandra Goble, Robert B. Adams, Dr. Julia V. Bukowski, Dr. Tony Frederickson, Rolf
Spiker, and Dr. Arnout C. Brombacher
~ William M. Goble

About the Authors

Harry L. Cheddie, P.Eng., CFSE

Harry Cheddie is Principal Engineer and Partner with exida. He is presently responsible for completing safety studies for end users, developing training programs, and teaching safety courses with an emphasis on IEC 61508 and IEC 61511.

Prior to joining exida, Harry was a Control Systems Advisor for Bayer Inc. in Sarnia, Ontario, Canada, where he was also the Supervisor of the Central Engineering Group responsible for process control systems design and maintenance.

Harry graduated from Salford University in the UK with a B.Sc. (1st Class Honors) degree in Electrical Engineering. He is a registered Professional Engineer in the province of Ontario, Canada.

Harry is certified by the American Society for Quality as a Quality Engineer, and as a Reliability Engineer. He is also a Certified Functional Safety Expert from the CFSE Governing Board.

Dr. William M. Goble, P.E., CFSE

William M. Goble is currently Principal Partner, exida, a company that does consulting, training and support for safety critical and high availability process automation. He has over 30 years of experience in control systems doing product development, engineering management, marketing, training and consulting.

Dr. Goble has developed several of the techniques used for quantitative analysis of safety (FMEDA) and reliability in automation systems. He teaches and consults in this area with instrumentation manufacturers and petrochemical/chemical companies around the world.

Dr. Goble has a BSEE from Penn State, an MSEE from Vilanova and a PhD from Eindhoven University of Technology in Reliability Engineering. He is a registereed professional engineer in the State of Pennsylvania and a Certified Functional Safety Expert (CFSE). He is a fellow member of ISA and previous author of the ISA book *Control Systems Safety Evaluation and Reliability*.

Contents

ABOUT THE AUTHORS vii

PREFACE xv

Chapter 1 THE SAFETY LIFECYCLE 1
Introduction, 1
Functional Safety, 1
Functional Safety Standards, 2
Safety Lifecycle, 5
Analysis Phase, 8
Realization Phase, 9
Operation Phase, 11
Benefits of the Safety Lifecycle, 12
Safety Lifecycle Adoption, 13
Exercises, 14
References and Bibliography 16

Chapter 2 SAFETY INSTRUMENTED SYSTEMS 19
Safety Instrumented Systems, 19
BPCS versus SIS, 20
SIS Engineering Requirements, 22
Safety Instrumented Function, 23
Exercises, 25
References and Bibliography 26

Chapter 3 EQUIPMENT FAILURE 27
Failure, 27
The Well-Designed System, 29
Failure Rate, 30

Time-Dependent Failure Rates, 30
Censored Data, 37
Confidence Factor, 38
Getting Failure Rate Data, 39
Exercises, 39
References and Bibliography 40

Chapter 4 BASIC RELIABILITY ENGINEERING 43
Measurements of Successful Operation – No Repair, 43
Useful Approximations, 49
Measurements of Successful Operation—Repairable Systems, 50
Periodic Restoration and Imperfect Testing, 56
Exercises, 57
References and Bibliography 59

Chapter 5 SYSTEM RELIABILITY ENGINEERING 61
Introduction, 61
System Model Building, 61
Reliability Block Diagrams, 62
Series System, 62
Parallel System, 63
Fault Trees, 65
Fault Tree Symbols, 67
Comparison of the Reliability Block Diagram and the Fault Tree, 68
Fault Tree AND Gates, 69
Fault Tree OR Gates, 69
Approximation Techniques, 69
Common Mistake, 71
Markov Models, 74
Markov Solution Techniques, 75
Realistic Safety Instrumented System Modeling, 78
Exercises, 79
References and Bibliography 81

Chapter 6 EQUIPMENT FAILURE MODES 83
Introduction, 83
Equipment Failure Modes, 83
Fail-Safe, 85
Fail-Danger, 85
Annunciation, 86
No Effect, 86
Detected/Undetected, 86
SIF Modeling of Failure Modes, 86
PFS/PFD, 87
PFDavg, 87
Exercises, 88

Chapter 7 SIF VERIFICATION PROCESS 89
The Conceptual Design Process, 89
Equipment Selection, 91
Redundancy, 94
SIF Testing Techniques, 96
Probabilistic Calculation Tools, 112
Verification Reports, 112
Exercises, 114
References and Bibliography 115

Chapter 8 GETTING FAILURE RATE DATA 117
Introduction, 117
Industry Failure Databases, 118
Product Specific Failure Data, 121
A Comparison of Failure Rates, 122
Comprehensive Failure Data Sources, 122
The Future of Failure Data, 122
Exercises, 126
References and Bibliography 127

Chapter 9 SIS SENSORS 129
Instrument Selection, 129
Diagnostic Annunciation, 129
Probabilistic Modeling of Sensors, 131
Pressure, 135
Temperature, 137
Level, 138
Flow, 140
Gas/Flame Detectors, 141
Burner Flame Detectors, 142
Miscellaneous, 143
Exercises, 143
References and Bibliography 144

Chapter 10 LOGIC SOLVERS 145
Introduction, 145
Relays/Pneumatic Logic, 145
Solid State / Intrinsically Safe Solid State, 146
Programmable Logic Controllers, 146
Safety Programmable Logic Controllers, 147
Probabilistic Modeling of the PLC, 150
Exercises, 154
References and Bibliography 154

Chapter 11 SIS FINAL ELEMENTS 157
Final Elements, 157
The "Well Designed" Remote Actuated Valve, 158
Actuator Types, 159
Valve Failure Modes, 160
Valve Types, 162
Probabilistic Modeling, 165
Failure Rate Comparison, 166
Diagnostics and Annunciation, 166
Exercises, 171
References and Bibliography 171

Chapter 12 TYPICAL SIF SOLUTIONS 173
Introduction, 173
Typical SIL 2 Architecture, 180
Some Common Hardware Issues Relating to the Various
    Solutions for SIL1, SIL2, and SIL3 Systems, 188
References and Bibliography 188

Chapter 13 OIL AND GAS PRODUCTION FACILITIES 189
Introduction, 189
Overall System Description, 191
Individual Well Controls and Shutdowns, 191
High Line Pressure Safety Instrumented Function (SIF), 195
SIF PFDavg Calculation, 196
Alternative SIF Designs, 209
Exercises, 214
References and Bibliography 214

Chapter 14 CHEMICAL INDUSTRY 215
Introduction, 215
Reactor, 215
Exercise, 228
References and Bibliography 228

Chapter 15 COMBINED BPCS / SIS DESIGNS 229
Introduction, 229
Analysis Tasks, 229
Alternative Designs, 231
Detailed Analysis of Combination BPCS and SIS Systems, 236
Exercises, 237
References and Bibliography 238

Appendix A STATISTICS 239

Appendix B   PROBABILITY   245

Appendix C   FAULT TREES   257

Appendix D   MARKOV MODELS   275

Appendix E   FAILURE MODES EFFECTS AND DIAGNOSTIC ANALYSIS (FMEDA)   303

Appendix F   SYSTEM ARCHITECTURES   315

Appendix G   MODELING THE REPAIR PROCESS   357

Appendix H   ANSWERS TO EXERCISES   367

INDEX   379

Preface

This book was written in response to many requests for more information regarding the details of probabilistic evaluation of safety instrumented functions. As the authors have had the great benefit of being asked to perform many such jobs, the problems apparent with previous methods have become clear to us.

This book would not have been written except for many individuals who have contributed with thought provoking questions and detailed comments on our work, our data and our methods. The authors would like to thank Dr. Julia V. Bukowski for many solutions to involved modeling questions. We thank Hal Thomas for many detailed questions and feedback on the initial answers. We thank Aart Pruysen, Rolf Spiker and Simon Brown for the detailed discussions on failure modes and interpretation.

We thank those who took the time to review our draft documents and provide valuable feedback. Those included Hal Thomas, Tim Layer, Vic Maggioli, Lindsey Bredemeyer, Eric Scharpf, Curt Miller, Oswaldo Moreno, Mike Bragg, Wally Baker and several others.

We thank ISA and particularly Lois Ferson who has guided us through the publishing process and shown great patience.

1

The Safety Lifecycle

Introduction

A working definition of the *Safety Lifecycle* is that it is an engineering process utilizing specific steps to ensure that Safety Instrumented Systems (SIS) are effective in their key mission of risk reduction as well as being cost effective over the life of the system. Activities associated with the Safety Lifecycle start when the conceptual design of facilities is complete and stop when the facilities are entirely decommissioned. Key activities associated with a Safety Lifecycle are outlined below.

- Analyzing risks
- Assessing the need for risk reduction
- Establishing system performance requirements
- Implementing the system according to the required performance criteria
- Assuring that the system is always correctly operated & maintained

Safety Lifecycle analyses heavily involve probabilistic calculations to verify the integrity of the safety design.

Functional Safety

The IEC 61508 standard defines safety as "freedom from unacceptable risk" (Ref. 1). Functional safety has been defined as "part of the overall safety relating to the process and the Basic Process Control System (BPCS) which depends on the correct functioning of the SIS and other protection layers." The phrase "correct functioning of the SIS" identifies the key concern. A high level of functional safety means that a safety instrumented system (SIS) will work correctly and with a high probability of success.

Functional safety is thus the primary objective in designing a safety instrumented system (SIS). To achieve an acceptable level of functional safety, several issues must be considered that may not be part of the normal design process for automation systems. These issues are provided as requirements in international standards.

Functional Safety Standards

As long as automated systems have existed, engineers have designed automatic protection into them. Early on, engineers often designed automatic protection systems using pneumatic logic or electrical relays because these components tend to fail in a de-energized mode. Systems were designed to be safe when the automation de-energized. They were, in other words, designed to "fail safe."

As the logic got more complicated, systems expanded to include large panels packed with relays and timers. In this same time frame of the '50s and '60s, printed circuit boards and solid-state electronics evolved. It was natural, therefore, for some more progressive engineers to convert control systems logic to new "solid-state" designs. Figure 1-1 shows an early solid-state module designed to implement burner logic. Unfortunately, there was little knowledge of the component failure modes of these designs.

When the first programmable electronic equipment, called programmable logic controllers (PLCs), were created as an alternative to relay logic, many engineers immediately believed these new devices would be perfect for automatic protection applications. They felt that the functionality of these electronic devices encompassed all that would be needed and more.

However, others realized that the failure characteristics of solid-state/programmable electronic equipment might be quite different from the traditional equipment. In fact, some government regulators banned the programmable electronic equipment outright. Regulatory persons began working with industry experts to establish guidelines for using electronic equipment in "emergency shutdown" applications. Eventually, international standards committees were formed, and standards covering the design and usage of equipment in "safety instrumented systems" (SIS) were published.

One of the more influential documents on SIS was called "Programmable Electronic Systems in Safety Related Applications," which was published by the Health and Safety Executive (HSE) in the United Kingdom (Ref. 2 and 3). Early national standards for SIS include "Grundsätze für Rechner in Systemen mit Sicherheitsaufgaben," published in Germany (Ref. 4 and 5) in 1990 and ANSI/ISA-84.01-1996, "Application of Safety Instrumented

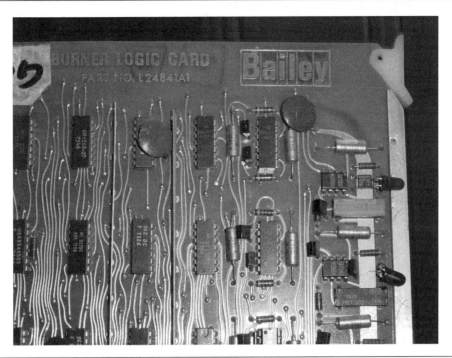
Figure 1-1. An Early Burner Logic Card Using Solid-State Components

Systems for the Process Industries" (Ref. 6) published in 1996 in the United States.

Many of the contributors to these national SIS efforts became members of an international committee that eventually wrote IEC 61508. This standard began in the mid 1980s when the International Electrotechnical Committee Advisory Committee on Safety (IEC ACOS) set up a task force to consider standardization issues raised by the use of programmable electronic systems (PES). Work began within IEC SC65A/Working Group 10 on a standard for PES used in safety-related systems. This group merged with Working Group 9, where a standard on software safety was in progress. The combined group treated safety as a system issue, and the first parts of IEC 61508 were published in 1998, with the release of final parts in 2000.

IEC 61508 is a basic safety publication of the IEC and as such, it is an "umbrella" document that covers multiple industries and applications. A primary objective of the standard is to serve as a guideline to help individual industries develop supplemental standards, tailored specifically for their industry but still in accordance with the original 61508 standard.

A secondary goal of the standard is to enable the development of electrical/electronic/programmable electronic (E/E/PE) safety-related systems where specific application sector standards do not already exist. IEC 61511 is an industry-specific standard for the process industries that is based on IEC 61508. ANSI/ISA-84.00.01-2004 (IEC 61511 Mod) was released as the USA version of IEC 61511 in September 2004. Note that it is identical to IEC 61511 in every detail, except it also includes a grandfather clause taken from OSHA 29CFR1910.119.

One clear goal of the committee was to create an engineering standard that would improve safety via the use of automation systems. Therefore, it was essential to understand the things that have gone wrong in the past. A study done by the Health and Safety Executive (Ref. 7) in the United Kingdom showed that many of the failures of automation systems to provide protection were caused by activities before and after the system had been designed. A graphic summary of the HSE study results is shown in Figure 1-2.

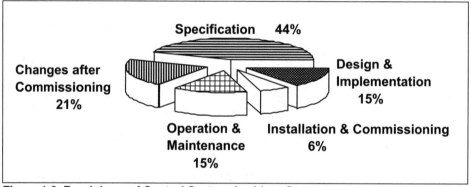

Figure 1-2. Breakdown of Control System Accident Causes

As indicated, a poor specification was the root cause of 44% of the automation system problems, which is by far the largest single cause. In effect, needed functionality was missing or incorrect. How can a control system designer create an automatic protection function when that designer does not know that it is needed?

The study found that many problems also occurred during installation, commissioning, operations, and maintenance. Largest of these were attributed to changes after commissioning (21%)—an understandably vulnerable aspect of automation systems.

In 2000 the US Environmental Protection Agency (EPA) and the Occupational Safety and Health Administration (OSHA) investigated recent accidents at chemical facilities and refineries (Ref. 8). When all the incidents were compared to one another, some common themes identified were:

1. **Inadequate hazard review or process hazards analysis** – In almost every accident, Process Hazards Analysis (PHA) was found to be lacking. This relates to identifying the hazards and properly

specifying the SIS based on the risk reduction required to mitigate hazardous events.

2. **Installation of pollution control equipment** – It was felt that this was also a reflection of inadequate hazards analysis and inadequate management of change procedures. This item was listed separately due to the large number of incidents of this nature.

3. **Use of inappropriate or poorly designed equipment** – For several accidents, the equipment used for a task was inappropriate. Better design performance was required.

4. **Warnings went unheeded** – Findings indicated that most incidents were often preceded by a series of smaller accidents, near misses, or accident precursors. Operations and maintenance procedures must include analysis, root cause investigation, and corrective action.

The EPA/OSHA findings support the results of the HSE study to a great extent. They highlighted that the key issues were identification of hazards and proper specification of the SIS taking the hazards into consideration. Both studies made it clear that focusing on programmable equipment and software design was not enough. A lifecycle approach was needed. In fact, a fundamental concept that has emerged from all functional safety standards is the use of the Safety Lifecycle.

Safety Lifecycle

The Safety Lifecycle (SLC) is an engineering process that contains all the steps needed to achieve high levels of functional safety during conception, design, operation, and maintenance of instrumentation systems. Its objective is clear! An automation system designed according to SLC requirements will predictably reduce risk in an industrial process. A simple version of the SLC is shown in Figure 1-3.

The safety lifecycle begins with the conceptual design of a process and ends only after the SIS is de-commissioned. The key idea here is that safety must be considered from the very inception of the conceptual process design and must be maintained during all design, operation, and maintenance activities. As indicated in the Figure 1-3 diagram, the safety lifecycle has three phases that can be identified as analysis (scope and hazard review portion of projects), realization (design portion of projects), and operation.

The safety lifecycle from the IEC 61508 standard is shown in Figure 1-4. This drawing provides more details of the safety lifecycle, but the three distinct phases are still clearly present. The IEC 61508 safety lifecycle shows that most of the activities of the analysis phase constitute a logical

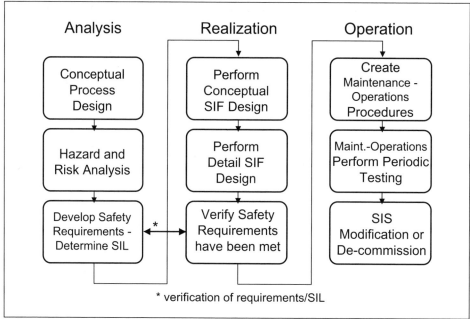

Figure 1-3. Simplified Safety Lifecycle Diagram

sequence. After hazard and risk analysis is performed safety requirements for the system are established.

Some safety requirements are met by external risk reduction facilities, including things like changes in process design, physical protection barriers, dikes, and emergency management plans. Some safety requirements are met by safety- related technology other than an SIS, including relief valves, rupture disks, alarms, and other specific-safety devices. Remaining safety functions are assigned as safety instrumented functions in a safety instrumented system.

The IEC 61508 safety lifecycle prescribes that planning for all maintenance testing and maintenance activities must be accomplished in the realization phase. To a certain extent this work can proceed in parallel with safety instrumented system design. This lifecycle diagram also shows that operation and maintenance responsibilities focus on periodic testing and inspection as well as management of modifications, retrofits, and eventual decommissioning.

An additional aspect regarding the IEC 61508 lifecycle is that there likely will be an overlap of the realization and operation activities with respect to steps 11 and 12.

The safety lifecycle from the ANSI/ISA84.00.01-2004 (IEC 61511 Mod) standard is shown in Figure 1-5. Although the drawing looks quite different from Figure 1-4, the fundamental requirements are much the same. Again, there are clear analysis, realization, and operation phases.

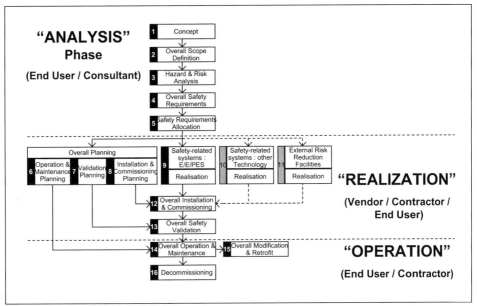

Figure 1-4. IEC 61508 Safety Lifecycle

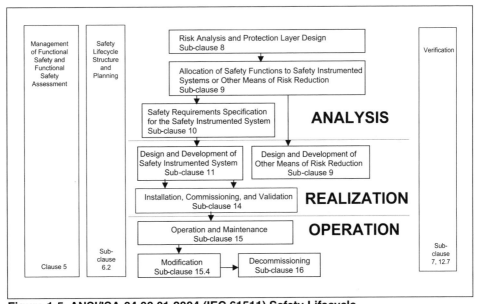

Figure 1-5. ANSI/ISA-84.00.01-2004 (IEC 61511) Safety Lifecycle

The safety lifecycle from ANSI/ISA-84.00.01-2004 (IEC 61511 Mod) was created specifically for the process industries, and thus many requirements are tailored for process applications. Note that there is an emphasis on managing functional safety; on the structure and planning of the safety lifecycle; and on verification throughout the entire lifecycle.

The IEC 61511 standard and the IEC 61508 standard emphasize good management of functional safety, which includes the application of a good plan and verification activities for each step of the safety lifecycle. These steps are shown as overtly drawn boxes on each side of the Figure 1-5 lifecycle diagram.

Analysis Phase

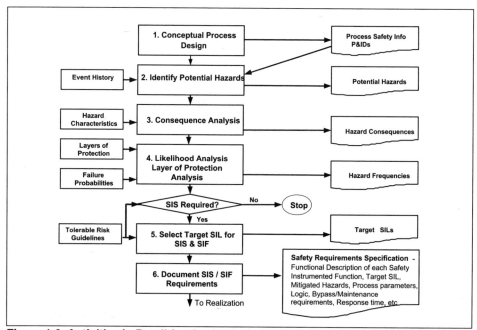

Figure 1-6. Activities in Detail for the Analysis Phase of the Safety Lifecycle

Figure 1-6 provides a detailed diagram of activities in the analysis phase of the safety lifecycle. When the conceptual process design is complete, detailed process safety information is available such as:

    a. types and quantities of chemicals used,

    b. pressures, temperatures, and flows of the intended design,

    c. process equipment used and its design strength,

    d. control strategy and intended control equipment,

    e. drawings, diagrams, and other relevant information.

Given this information, a hazard and risk analysis is performed (Ref. 9, 10, and 11) that identifies possible hazards and then establishes the consequence and likelihood of each. On some projects, consequence is determined through detailed analysis, and on others it is done by estimation. Likewise, likelihood analysis is sometimes performed by detailed analysis and sometimes by estimation. With the emergence of new techniques such as Layer of Protection Analysis; however, the trend is clearly toward more analysis (Ref. 12 and 13).

The consequence severity and the likelihood frequencies determine risk. In some cases, the risk of a hazard is within tolerable levels, and no risk reduction is needed. For these cases, no SIS is required. In other cases, risk reduction is required, and the quantity of risk reduction is specified by an order-of-magnitude level called the safety integrity level (SIL) as indicated in Figure 1-7.

Safety Integrity Level	Probability of Failure on Demand (PFDavg.) Low Demand Mode	Risk Reduction Factor (RRF)
4	$10^{-4} > PFDavg \geq 10^{-5}$	$10000 \leq RRF < 100000$
3	$10^{-3} > PFDavg \geq 10^{-4}$	$1000 \leq RRF < 10000$
2	$10^{-2} > PFDavg \geq 10^{-3}$	$100 \leq RRF < 1000$
1	$10^{-1} > PFDavg \geq 10^{-2}$	$10 \leq RRF < 100$

Figure 1-7. A Chart of Safety Integrity Levels (SIL)

Each safety instrumented function is documented in the safety requirements specification. That document (or collection of documents) includes all functional information, logic, performance information, timing, bypass/maintenance requirements, reset requirements, the safety integrity level for each safety instrumented function, and any other requirement information that the designers may need.

Realization Phase

When all safety instrumented functions are identified and documented, the design work can begin (Figure 1-8). A conceptual design is performed by choosing the desired technology for the sensor, the logic solver, and the final element. Redundancy may be included so as to achieve high levels of safety integrity, to minimize false trips, or for both reasons.

Once the technology and architecture have been chosen, the designers review the periodic test philosophy constraints provided in the SRS. Given that safety instrumented systems will, hopefully, not be called on to activate, they must be completely inspected and tested at specified time intervals. This periodic testing is performed to ensure that all the elements of the system are fully operational and to verify that no failures have occurred.

In some industries, the target periodic test interval corresponds with a major maintenance cycle, for example, two, three, or even five years. In other industries, a periodic inspection must be done more frequently. If these tests must be performed while the process is operating, online test facilities are designed into the system. A periodic inspection and test plan is required for all the instrumentation equipment in each safety instrumented function.

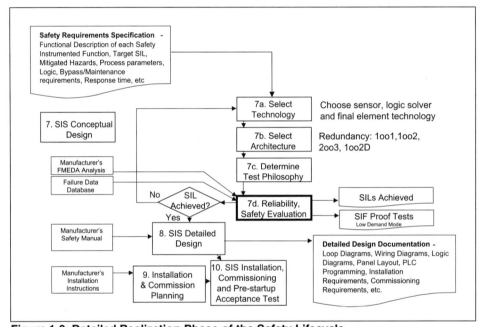

Figure 1-8. Detailed Realization Phase of the Safety Lifecycle

Once the technology, architecture, and periodic test intervals are defined, the designers do a reliability and safety evaluation (Ref. 14 and 15) to verify that the design has met the target safety integrity level and reliability requirements. In the past, this probabilistic evaluation has not been part of a conventional design process. The effort requires gathering failure rate data as a function of failure modes for each piece of equipment in the safety instrumented function.

Results of the evaluation typically include a number of safety integrity and availability measurements. Most important, the average probability of failure on demand (PFDavg) and the safe failure fraction (SFF) is calculated for low demand mode. Probability of failure per hour is calculated for high demand mode. From charts, the SIL level that the design achieves is known. Many designs are not sufficient, and the designers repeat design steps until an optimal design is attained.

When the best conceptual design is complete, detail design activities can begin. These activities include much of the normal project engineering that is performed by integration companies and project engineering teams. The manufacturer's safety manual must be consulted for each piece of equipment in the design in order to ensure that all manufacturers' safety restrictions are being followed.

A plan is created for the installation and commissioning. This step includes a comprehensive test to validate that all requirements from the original SRS have been completely and accurately implemented. Also needed is a revalidation plan that is a subset of the validation plan. When the installed system is tested and validated, the safety instrumented system is ready to provide protection when actual operation begins.

Operation Phase

The operation phase (Figure 1-9) of the safety lifecycle begins with a pre-startup safety review (PSSR) of the SIS design. During the PSSR, the engineers must answer a number of questions. Does the system meet all the requirements stated in the safety requirements specification? Have all safety instrumented functions met SIL targets and Mean Time To Trip Spuriously (MTTFS) targets? Have all the necessary design steps of the safety lifecycle been carried out successfully? Has the manufacturer's safety manual been followed for all equipment? Is there a periodic inspection and test plan for each safety instrumented function? Have the maintenance procedures been created and verified? Is there a management-of-change procedure in place? Are operators and maintenance personnel qualified and trained? If the answers to these questions are acceptable, the process can proceed with startup and operation.

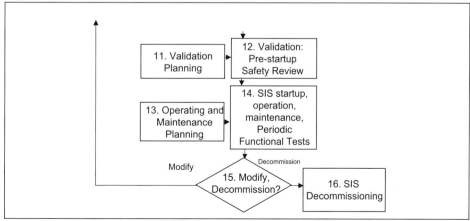

Figure 1-9. Detailed Operation Phase of the Safety Lifecycle

While in operation, proper operating procedures and all maintenance activities including periodic function testing and mean time to repair targets have to be followed. Periodic functional testing must be done as per the time schedule established during the conceptual design verification calculations and must be done per the plan established to ensure that all potentially hidden dangerous failures are detected. All periodic inspection and test activities must be documented. It is very important that each safety instrumented function is restored to full operation after each test. Bypass valves and force functions must be removed and these restorations must be documented.

The safety lifecycle includes management of all modifications made to the system during its useful life. For each, the engineer making the change must analyze the impact of the change and go back to the appropriate step in the safety lifecycle. If new technology is chosen, the SIL verification must be repeated. The new SIL level must meet or exceed the original.

Decommissioning is considered as well. The engineer must analyze the effect of decommissioning the system. Are all safety instrumented functions no longer needed? If some are still needed, they must be relocated or decommissioning must not proceed.

Benefits of the Safety Lifecycle

The safety lifecycle was created not only to help designers of safety instrumented systems build safer systems, but also to help create more cost-effective systems. The process is essentially a feedback system incorporating two important verifications. The first is after conceptual design and the second is ongoing periodically during operation, as shown in Figure 1-10.

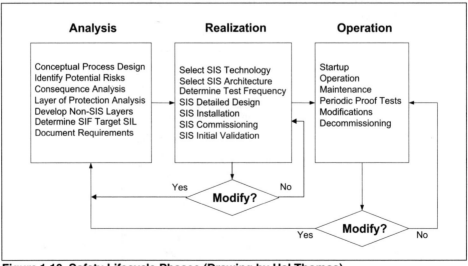

Figure 1-10. Safety Lifecycle Phases (Drawing by Hal Thomas)

By analyzing the process risk, a system can be designed to enable that particular risk to be reduced to a tolerable level—not over-designed and not under-designed. Costly problems of the past included systems that were over-designed so as to meet older "prescriptive" standards. Previous systems have also been designed that would not provide an expected risk reduction because the designers did not understand the performance capabilities relative to failure modes of equipment being used.

The probabilistic verification included in the safety lifecycle helps engineers create "balanced" designs (Ref. 16). In the past, designers working on a SIL3 safety problem were focused on the logic solver. Best practices indicated that a SIL3 safety-certified redundant logic solver should be used. This solver may have been used in combination with sensing switches and simple mechanical final elements. An analysis of such a design will show that, by today's standards, it may only meet SIL1 with frequent periodic tests and inspections. Balanced designs put equal emphasis on all parts of the safety instrument function — logic solvers and field devices.

The performance-based nature of the probabilistic SIL verification also allows designers the freedom to innovate, to find better designs that meet safety integrity performance, and to reduce lifecycle cost.

Perhaps most important, the safety lifecycle concept as presented in international standards provides a common way to design throughout the world. No longer are there specific national standards with different details that change from country to country — at least for the process industries.

Safety Lifecycle Adoption

Although the safety lifecycle was presented in the standard ANSI/ISA-84.01 in 1996, many companies were focused on other problem areas, including the "Y2K" threat. All process industry companies did not adopt the ISA-84.01 safety lifecycle right away. In 2000, the United States Occupational Safety and Health Administration (OSHA) responded to ISA questions in a letter (Ref. 17), which stated that "As ISA-84.01 is a national consensus standard, OSHA considers it to be a recognized and generally accepted good engineering practice for SIS." This letter generated significant discussion and, because it appeared right after the Y2K crisis, the safety lifecycle concept began to receive attention.

The final parts of IEC 61508 were published in February 2000 and these in combination with the OSHA letter in the United States generated a great deal of attention. Interest around the world also grew as IEC 61508 became the enforcement standard of other regulatory bodies. Regulators in British Commonwealth countries began referring to IEC 61508 even before it was released.

Surveys indicate that an increasing number of companies worldwide are modifying their company engineering procedures to reference IEC 61508 or IEC 61511. One estimate in 2003 indicated that 70 percent of the respondents in the petrochemical and chemical industries were "using or planning on using" the safety lifecycle.

It must be recognized that the Safety Lifecycle and associated Safety Instrumented Systems need to be part of an overall Process Safety Management System (PSM) for the entire plant. PSM can be defined as a program or set of activities involving the application of management principles and analyses to ensure the overall process safety of process plants. PSM, therefore, covers all aspects of process safety, not just functional safety.

For the Safety Lifecycle to be properly adopted and implemented, it must be integrated into the overall PSM for the plant so that its importance is well recognized and prioritized. This integration is sometimes the biggest challenge for a plant due to organization barriers.

Exercises

1-1. International Safety Standards require that operating companies follow the safety lifecycle specifically as outlined in the respective standard.

   a. True
   b. False

1-2. The safety lifecycle process can:

   a. reduce SIS costs.
   b. increase process safety.
   c. help ensure that regulations are met.
   d. provide an example of "good engineering practices."
   e. All of the above

1-3. Why should a company pay attention to IEC 61508 and IEC 61511

   a. It is legally required in some countries.
   b. It can save money on safety systems.
   c. Owner/operators often require compliance.
   d. A combination of answers a, b, and c

1-4. According to the safety lifecycle in IEC 61508, when should a process hazards analysis be conducted?

   a. Immediately after defining the project scope

   b. Immediately before specifying the overall safety requirements

   c. Immediately before verifying the Safety Instrumented Function (SIF) will achieve the required risk reduction

   d. Both a and b are correct.

1-5. In the safety lifecycle, a safety requirements specification is done after:

   a. defining the project scope.

   b. the Hazard and Risk Analysis Phase.

   c. throughout the phases of the lifecycle.

   d. overall Safety Validation.

1-6. Which of the following methods is **not** usually part of the analysis phases of the safety life cycle?

   a. Layer of Protection Analysis (LOPA)

   b. SIL Verification Analysis

   c. HAZOP

   d. Risk Analysis

1-7. What parts of the safety lifecycle are subject to functional safety management?

   I. The analysis phase
   II. The operation phase
   III. Only the phases before commissioning
   IV. The design phase.

   a. Only I

   b. Only I and III

   c. Only I, III, and IV

   d. Only I, II, and IV

1-8. In the analysis phase of the safety lifecycle, the following activity is done:
   a. Install SIS
   b. Analyze Risk
   c. Decommission SIS
   d. Operation and validation planning
   e. SIS validation

1-9. When does the safety life cycle end?
   a. It never ends.
   b. When the project is fully commissioned
   c. When the safety system is decommissioned
   d. When the safety system is proven in use

REFERENCES AND BIBLIOGRAPHY

1. IEC 61508, *Functional Safety of Electrical / Electronic / Programmable Electronic Safety-related Systems*, Geneva: Switzerland, 2000.

2. *Programmable Electronic Systems in Safety Related Applications, Part 1, An Introductory Guide*, U.K.: Sheffield, Heath and Safety Executive, 1987.

3. *Programmable Electronic Systems in Safety Related Applications, Part 2, General Technical Guidelines*, U.K.: Sheffield, Heath and Safety Executive, 1987.

4. DIN V VDE 0801, *Grundsätze für Rechner in Systemen mit Sicherheitsaufgaben*, 1990.

5. DIN V VDE 0801 A1, *Grundsätze für Rechner in Systemen mit Sicherheitsaufgaben, Änderung A1*, 1994.

6. ANSI/ISA-84.01-1996, *Application of Safety Instrumented Systems for the Process Industries*, NC: Research Triangle Park, ISA, 1996.

7. *Out of Control: Why Control Systems Go Wrong and How to Prevent Failure*, U.K.: Sheffield, Heath and Safety Executive, 1995.

8. U.S. Environmental Protection Agency and United States Occupational Safety and Health Administration, *EPA/OSHA Joint Chemical Accident Investigation Report: Recurring Causes of Recent Chemical Accidents*, 2000

9. Henley, E. J., and Kumamoto, H., *Probabilistic Risk Assessment: Reliability Engineering, Design, and Analysis*, NJ: Piscataway, IEEE Press, 1992.

10. *Guidelines for Safe Automation of Chemical Processes*, Center for Chemical Process Safety, American Institute of Chemical Engineers, NY: New York, 1993.

11. Lees, F. P., *Loss Prevention for the Process Industries*, U.K.: London, Butterworth and Heinemann, 1992.

12. *Layer of Protection Analysis*, Center for Chemical Process Safety of the American Institute of Chemical Engineers, NY: New York, 2003.

13. Marszal, Edward M., and Scharpf, Eric W., *Safety Integrity Level Selection Including Layer of Protection Analysis*, NC: Research Triangle Park, ISA, 2003.

14. Goble, W. M., *Control Systems Safety Evaluation and Reliability*, second edition, NC: Research Triangle Park, ISA, 1998.

15. TR84.00.02-2002, Technical Report, *Safety Instrumented System (SIS) – Safety Integrity Level (SIL) Evaluation Techniques*, NC: Research Triangle Park, ISA, 2002.

16. Gruhn, P., "Safety Systems, Peering Past the Hype," *1996 Proceedings of the 51st Instrumentation Symposium for the Process Industries*, NC: Research Triangle Park, ISA, 1996.

17. Letter to Ms. Lois M. Ferson from OSHA, March 23, 2000, D.C.: Washington, Occupational Safety and Health Administration, Reply to DCP/GICA/MLMG-1356, 2000.

2
Safety Instrumented Systems

Safety Instrumented Systems

The ANSI/ISA-84.00.01-2004 (IEC 61511) standard (Ref. 1) defines a safety instrumented system (SIS) as an "instrumented system used to implement one or more safety instrumented functions. A SIS is composed of any combination of sensor(s), logic solver(s), and final element(s)." IEC 61508 (Ref. 2) does not use the term SIS but instead uses the term "safety-related system." That term defines the same concept but uses language that can be broadly applied to many industries.

Practitioners often prefer a more functional definition of SIS such as: "A SIS is defined as a system composed of sensors, logic solvers and final elements (Figure 2-1) designed for the purpose of:

- automatically taking an industrial process to a safe state when specified conditions are violated;

- permit a process to move forward in a safe manner when specified conditions allow (permissive functions); or

- taking action to mitigate the consequences of an industrial hazard."

The definition tells us that a SIS may be responsible for shutdown functions, permissive functions, and even consequence reduction (mitigation) functions. All these functions have a common attribute — they all reduce risk. One common interpretation of a SIS definition is, therefore, "automatic risk reduction systems." In some cases the function is designed to reduce risk by decreasing the likelihood of a potential hazard. In some cases the function will decrease risk by reducing the magnitude of the consequence.

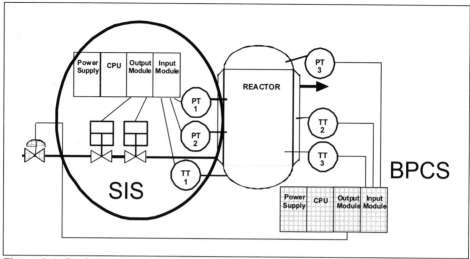

Figure 2-1. Basic Process Control System versus Safety Instrumented System

BPCS versus SIS

A safety instrumented system, like a basic process control system, is also composed of sensors, controller(s), and final elements. Although much of the hardware appears to be similar, safety instrumented systems and basic process control systems differ very much in function. The primary function of a control loop is generally to maintain a process variable within prescribed limits. A safety instrumented system monitors a process variable and initiates action when required.

Even with the described functional differences, when one reads the definition of a SIS one naturally asks a few questions. "What is so special about a SIS?" "What is different between a SIS and a BPCS?" Many aspects are the same.

The engineering skills needed to design a good control system are needed to design a good safety system. Detailed knowledge of how the instrumentation works applies to both types of systems. Knowledge of the materials of construction and process material compatibility is essential to successful design in both types of systems.

The commonality of a SIS and a BPCS has led many to treat the design process for each the same. Some even use their control equipment to perform the safety instrumented functions.

The SIS operates with completely different dynamics, however. Remember that the process is designed to be safe under normal conditions and process designers work hard to minimize process hazards within the process itself.

When the process design is successfully oriented toward safety, risk analysis teams will usually estimate residual hazardous situations that occur only once in many years (clearly low demand mode). The average time period between hazardous events is often estimated to be over 10 years. Thus, we have a situation where the SIS is activated only once every ten years or more. During normal operation it is static. A safety isolation valve may sit motionless for years! Contrast this with the operation of a BPCS. The control signals are normally dynamic with some signals moving considerably at all times.

BPCS Failure Modes	SIS Failure Modes
Control Output Saturated High Control Output High Control Output Frozen Control Output Low Control Output Saturated Low Control Output Slow to Respond Control Output Too Fast Control Output Erratic Process Variable Indication Saturated High Process Variable Indication High Process Variable Indication Frozen Process Variable Indication Saturated Low Process Variable Indication Low Control Output Indication Saturated High Control Output Indication Frozen Control Output Indication Saturated Low Process Variable Indication Erratic	Fail to Function (Fail – Danger) Spuriously Function (Fail – Safe) Function Delayed (Fail - Danger)

Figure 2-2. Failure Mode Comparison of BPCS and SIS

The AIChE (American Institute of Chemical Engineers) CCPS (Center for Chemical Process Safety) Process Equipment Reliability Database (PERD) initiative has rigorously identified and documented failure modes for instrument loops, which encompass control, indication, alarm, and automatic protection (Ref. 3). An excerpt from those lists shows a comparison of failure modes applicable to BPCS versus SIS (Figure 2-2). The ability of operations personnel to detect these failure modes is also quite different.

Remember that the BPCS operates with signals that are relatively dynamic. This makes BPCS failures generally detectable by plant personnel. Example diagnostic methods include flat line outputs, quality indicators, pre-alarms, deviation alarms, and out of range signals. While a BPCS operates under relatively dynamic conditions, safety instrumented system signals are static Boolean variables. Since the SIS only takes action when a potentially dangerous condition is detected, it can be very hard for operations and maintenance personnel to detect certain failure modes of a SIS.

SIS Operating Condition	Process	Protection Available	Failure Indication
Normal	Operating Normally	YES	Not Applicable
Fail-Safe	Falsely Shutdown	Not Applicable	YES
Fail-Danger	Operating Normally	NO	Not Without Diagnostics

Figure 2-3. SIS Operating Conditions

Figure 2-3 shows a chart of SIS operating conditions. The fail-danger mode is the fundamental problem. Under those conditions the process is operating normally but without the automatic protection of the SIS. Without special diagnostics, operations personnel have no indication that something has failed.

SIS Engineering Requirements

While a SIS is similar to a BPCS in many ways, the differences result in unique design, maintenance, and mechanical integrity requirements. These include:

- Design to fail-safe
- Design diagnostics to automatically detect fail-danger
- Design manual test procedures to detect fail-danger
- Design to meet international and local standards

Remember that the design engineer must satisfy these requirements in addition to the functional requirements and performance requirements normally associated with control system design.

Another difference between Basic Process Control System design and Safety Instrumented System design is that per ANSI/ISA-84.00.01-2004 (IEC 61511 Mod) these systems are designed and implemented to meet different risk reduction requirements presented by the various hazards. (Chapter 1)

So while it may seem like BPCS design and SIS design are the same, most experienced SIS designers recognize the additional design constraints that must be met for SIS.

Safety Instrumented Function

In ANSI/ISA-84.00.01-2004 (IEC 61511 Mod), 3.2.71, a safety instrumented function is defined as a "safety function with a specified safety integrity level which is necessary to achieve functional safety." This standard, 3.2.68, defines a safety function as a "function to be implemented by a SIS, other technology safety-related system or external risk reduction facilities, which is intended to achieve or maintain a safe state for the process, with respect to a specific hazardous event."

Examples of potential safety instrumented functions include the following:

- Close outlet valve in a separation unit to prevent high pressure from going downstream, which might result in vessel rupture and explosion.

- Cut off fuel flow in an industrial burner when fuel pressure is too low to sustain combustion, which might result in flameout and possible explosion due to fuel buildup in the combustion chamber.

- Open coolant flow valve to prevent column rupture due to over temperature.

- Close connection valve to isolate reactants to prevent unit overpressure when reverse flow detected.

- Close valve to stop material flow into a tank to prevent spillage if high level is detected, which might result in environmental damage.

- Open sprinkler valve when a flame is detected in order to reduce the size of a fire.

Equipment Used in a Safety Instrumented Function

Emphasis should be placed on the last phrase of the SIF definition, "specific hazardous event." This phrase helps one clearly identify what equipment is included in the safety instrumented function versus auxiliary equipment not actually needed to provide protection against the hazard.

Consider the example of a safety instrumented function to protect against vessel rupture due to over-pressure. When high pressure above the trip point is detected, the function will do three things:

1. It will close a valve to stop material flow into a process unit.
2. It will turn off the pump used for this material.
3. It will close an outlet isolation valve to isolate the unit from the remainder of the plant.

It seems logical to list the following equipment for this safety instrumented function: pressure transmitter, logic solver, inlet feed valve, pump control relay, and outlet isolation valve. However, for each piece of equipment ask the question, "Is this piece of equipment needed to protect against the specific hazardous event?" In this case, the pump is turned off just to protect the pump from backpressure burnout. (NOTE - This may be part of another safety instrumented function.) The outlet isolation valve is closed in order to avoid process disruptions in the remainder of the plant. Neither is required to protect against the hazard and should not be included in the SIF verification calculation. The pump control relay may be part of another SIF intended to protect the pump. However, it is likely that this SIF may have a lower safety integrity requirement.

Personnel as Part of a Safety Instrumented Function

Many never consider a person to be part of a safety instrumented function. When reviewing the definitions used in functional safety standards, there is little indication that this was ever intended. A process operator takes action when normal process operating conditions are violated. The operator normally responds to these violations from alarms. The operator action is, therefore, normally considered as being part of the alarm layer of protection, not the SIS.

For this reason SIL determination for operator action is normally not completed. However, a note in ANSI/ISA-84.00.01-2004 (IEC 61511 Mod) (3.2.72, Note 4) states, "When a human action is a part of an SIS, the availability and reliability of the operator action must be specified in the SRS and included in the performance calculations for the SIS." This implies a broad definition of SIF that includes a human action.

Exercises

2-1. Compared to control system design, there are special considerations and additional requirements for SIS design including:

   I. the need to meet international standards.
   II. careful consideration of system failure modes.
   III. the physical design of the system.
   IV. the ability of the system to be located in an environment with flammable material.

   a. I and II
   b. I and IV
   c. III and IV
   d. II and IV

2-2. A safety instrumented system typically performs the following functions:

   I. reads sensors and performs optimization calculations.
   II. generates outputs to maintain process variables close to setpoints.
   III. executes pre-programmed actions to avoid or mitigate process hazards.

   a. I and III
   b. II and III
   c. III
   d. I, II, and III

2-3. A safety instrumented system has to include:

   a. sensors.
   b. logic solvers.
   c. final elements.
   d. all of the above.

2-4. In a de-energize to trip system, a loss of power supply failure is considered

   a. impossible.
   b. fail-danger.
   c. not a failure.
   d. fail-safe.

2-5. How does a Safety Instrumented Function most typically reduce risk?

   a. Reduces the likelihood of harm.

   b. Reduces the magnitude of harm.

   c. Satisfies legal requirements.

   d. Satisfies managerial requirements.

2-6. Which safety standards address SIL selection methods?

   I. IEC 61508
   II. IEC 61511
   III. ISO 9000
   IV. ANSI/ISA84.00.01-2004

   a. I and IV only

   b. I, II, and IV only

   c. All four items

   d. I and II only

2-7. Should a Fire and Gas system that performs a mitigation function be classified as a Safety Instrumented System?

   a. Yes – it reduces risk.

   b. No – these systems are not classified as a SIS.

2-8. A cause and effect diagram shows logic for a function where two valves will be closed and a pump will be de-energized when either of two pressure sensors senses that pressure in a vessel goes too high. Valve 1 cuts off inlet feed to reduce pressure in the vessel. Valve 2 closes the outlet to prevent process disturbance downstream. De-energizing motor contactors turn off the pump in order to prevent pump failure. What equipment is part of the safety instrumented function?

REFERENCES AND BIBLIOGRAPHY

1. ANSI/ISA-84.00.01-2004, *Functional Safety: Safety Instrumented Systems for the Process Industry Sector – Parts 1, 2, and 3 (IEC 61511 Mod)*, NC: Research Triangle Park, ISA, 1996.

2. IEC 61508, *Functional Safety of electrical / electronic / programmable electronic safety-related systems*, Geneva: Switzerland, 2000.

3. Arner, D. C., and Angstadt, W. C., "Where For Art Thou Failure Rate Data," *Proceedings for ISA 2001 – Houston*, NC: Research Triangle Park, ISA, 2001.

3
Equipment Failure

Failure

A failure occurs when a device at some level (a system, a unit, a module, or a component) fails to perform its intended function. To many, the definition is clear. Disagreement may occur, but when this happens it is usually a matter of properly defining "intended function." For safety instrumented systems, the definition of intended function is usually clear and should be properly recorded in the safety requirements specification.

Each safety instrumented function (SIF) in a safety instrumented system must perform its protection function, must not falsely shut down the process, and must perform ancillary functions such as communications and diagnostics. However, all these functions are not necessarily included in all analyses. Diagnostic functions or communications functions may or may not be counted as a failure depending on the purpose of the analysis.

What if certain components are used only for diagnostic purposes and they fail? The safety protection functionality will continue to work perfectly but the diagnostic function no longer works. Is this failure considered when calculating probability of failure on demand? Generally not! This is justified in many cases because the safety protection function will operate even when the diagnostics do not.

However, as we will see, if the safety instrumented function verification takes credit for the diagnostics to achieve a sufficiently high level of safety integrity, then the diagnostic failures will need to be modeled in an accurate probabilistic SIF verification.

Random Failures vs. Systematic Failures

Two fundamentally different categories of failures exist: physical failures (often called random failures) and functional failures (often called systematic failures). (See Figure 3-1). Random failures are relatively well understood. A random failure is almost always permanent and attributable to some component or module. For example, a system that consists of a programmable electronic controller module fails. The controller output de-energizes and no longer supplies current to a solenoid valve. The controller diagnostics identify a bad output transistor component.

An "autopsy" of the output transistor shows that it would not conduct electrical current and failed with an open circuit. The failure occurred because a thin bond wire inside the transistor melted. Plant operations reports a nearby lightning strike just before the controller module failed. An electrical surge would cause an impulse of electrical stress that can exceed the strength of a transistor.

It should be noted that lightning is considered a random event. Many failure classification schemes use the term *random failure* because stress events are generally random. Some failure classification schemes use the term *physical failure* for the same thing.

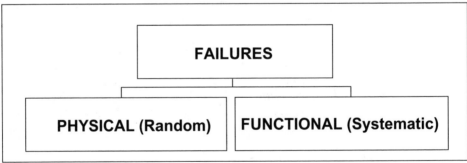

Figure 3-1. Random versus Systematic Failures

In another example of a physical failure, a system failed because a power supply module failed with no output. An examination of the power supply showed that it had an open electrolytic capacitor (a capacitor with a value far below its rating, which is virtually the same as an open circuit). The capacitor autopsy showed that the electrolyte had evaporated. Without electrolyte, these capacitors cannot hold a charge and become open circuits. Evaporation is a process that occurs over time. Eventually, all electrolyte in a electrolytic capacitor will evaporate and the capacitor ceases to perform.

Other failures are called systematic failures. A systematic failure occurs when the system is capable of operating but does not perform its intended function. An example of this is the common "software crash." A piece of data is entered into a personal computer and, when the return key is pressed, the computer simply quits. The computer has failed to accomplish its intended function, but no component has failed — no random failure has occurred. The computer re-starts properly after the "reset" button is pressed. The failure may or may not appear to be repeatable depending on data entered.

If an engineer programming a safety function entered an incorrect logic block such that a safety instrumented function would not perform its protective function, that failure would also be considered a systematic failure. Again, the hardware is fully capable of executing the programmed logic, no random failure has occurred, but the safety instrumented function would not work.

Systematic failure sources are almost always design faults, usually due to inadequate procedures or training. Occasionally however, a maintenance error or an installation error causes a systematic failure. The exact source of a systematic failure can be obscure. Often the failure will occur when the system is asked to perform some unusual function, or perhaps the system receives some combination of input data that was never tested.

Some of the most obscure failures involve combinations of stored information, elapsed time, input data, and function performed. Systematic failures may be permanent or may be transient in nature. Up to the present, failure rate data for systematic failures has been difficult to acquire. However, possibilities exist for better data acquisition in the future.

The Well-Designed System

Current functional safety standards, IEC 61508 and ANSI/ISA-84.00.01-2004 (IEC 61511 Mod), (Ref. 1 and 2) state that probabilistic evaluation using failure rate data be done only for random failures. To reduce the chance of systematic failures, the standards include a series of "design rules" in the form of specific requirements. These requirements state that the safety instrumented system designer must check a wide range of things in order to detect and eliminate systematic failures.

Examples of these requirements include the need to check that an instrument is being used within the stated environmental conditions and application restrictions of the manufacturer. Specific requirements are made to check for compatible materials whenever any part in a safety instrumented function has process wetted parts. A planned comprehensive testing program with written test results is required. A periodic inspection and test plan must be created and verified.

A "process safety review" is done before startup that includes a second check that all other requirements have been met. Systems designed to meet all these requirements should have a sufficiently low level of systematic failures. These systems have been referred to as "well-designed systems."

Failure Rate

One of the basic concepts of functional safety standards is the use of probabilistic evaluation of random hardware failures in order to verify that the SIF design meets the safety integrity requirements established during the process hazards analysis. This evaluation is based upon principles developed over a period of several decades in the field of reliability engineering. Reliability engineers gather statistics about when and how things fail. This information is used to predict future performance of SIF designs.

In reliability engineering, the primary statistical variable of interest is *Time To Failure (T)*. The time to failure measurement can be analyzed to generate another important measurement, failure rate. Instantaneous failure rate is a commonly used measure of reliability that gives the number of failures per unit time from a quantity of components exposed to failure.

$$\lambda(t) = \text{Failures per Unit Time/Quantity Exposed} \tag{3-1}$$

A failure rate is typically represented by the lower case Greek letter lambda (λ). A failure rate has units of inverse time. In the design of electronic devices, it is a common practice to use units of "failures per billion (10^9) hours." This unit is known as FIT for Failure unIT. For example, a particular integrated circuit will experience seven failures per billion operating hours at twenty-five degrees C and, thus, has a failure rate of seven FITs. It is also common to use units of "failures per million hours" or "failures per year."

EXAMPLE 3-1

Problem: A failure rate is given in units of 3200 FITS. What is the failure rate in units of failures per year?

Solution: 0.0000032 failures per hour × 8760 hours per year = 0.028032 failures per year.

Time-Dependent Failure Rates

If an analyst is given specific time-to-fail data for a collection of components, the instantaneous failure rate can be calculated. For example, if thirty power supplies were tested for time to failure, the data could be similar to that listed in Table 3-1. (NOTE: This is an example only, not real data.)

The thirty units are placed into a stressful environment (likely an oven) to accelerate failure mechanisms. The first unit failed after thirty-three hours. The failure rate can be calculated from the equation:

$$\lambda = \Delta Nf/(Ns \times \Delta t) \qquad (3\text{-}2)$$

where:
- ΔNf = the number of failed units during the time period
- Ns = the number of surviving units at the end of the time period
- Δt = the time period between failures

Since the first unit failed after thirty-three hours:
$\Delta Nf = 1$
$Ns = 29$ (30 units started the test and one failed, therefore 29 remain operating at the end of the time interval)

and
$\Delta t = (33 - 0) = 33$

For the first time interval, the failure rate equals 0.001045 failures per hour.

If the second power supply fails after 96 hours then:
$\Delta Nf = 1$
$Ns = 28$

and
$\Delta t = (96 - 33) = 63$

For the second time interval the failure rate equals 0.000567 failures per hour.

Table 3-1 shows the failure times for all units and a calculation of the failure rate during the operating time interval of the set. If the failure rate is plotted as a function of the operating time interval, the shape of the curve is shown in Figure 3-2.

The shape of the plot in Figure 3-2 is characteristic of many components and well known to reliability engineers. The shape is called the "bathtub curve." Three regions are distinct. In the early portion of the plot, failure rates are higher. This area is called "infant mortality." The middle portion of the curve is known as "useful life." The final portion of the curve is called "end of life" or "wearout region."

Table 3-1. Time to Failure Data

Failure Number	Hours	Failure Rate
1	33	0.001045
2	96	0.000567
3	196	0.000370
4	240	0.000874
5	409	0.000237
6	614	0.000203
7	831	0.000200
8	1045	0.000212
9	1282	0.000201
10	1540	0.000194
11	1815	0.000191
12	2106	0.000191
13	2414	0.000191
14	2740	0.000192
15	3091	0.000190
16	3471	0.000188
17	3886	0.000185
18	4348	0.000180
19	4862	0.000177
20	5431	0.000176
21	6056	0.000178
22	7499	0.000087
23	8339	0.000170
24	9270	0.000179
25	10305	0.000193
26	11460	0.000216
27	12751	0.000258
28	13351	0.000833
29	13853	0.001992
30	13990	

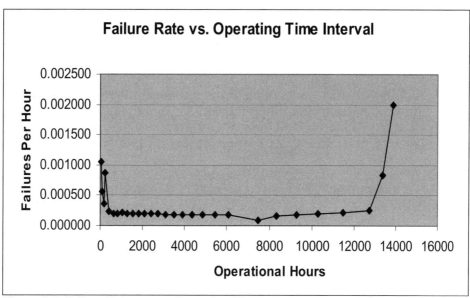

Figure 3-2. Failure Rate versus Operating Time

Failures occur when some "stress" exceeds the corresponding "susceptibility" in any component. As a concept it makes perfect sense in the context of mechanical engineering. The stress is usually a force, and the susceptibility is the point where the mechanical component can no longer resist that force. An analogous concept has been developed for electronic components (Ref. 3). Simulations using the stress-susceptibility concept generate failure rate curves similar to the bathtub curve (Ref. 4 and 5).

In the early portion of the bathtub the failure rate is higher because weaker units in the population will fail first. As these weaker units are removed from the population, the failure rate goes down. This often occurs over a period of time involving a few weeks. A considerable effort has gone into the creation of procedures to identify and weed out weak units in manufacturing processes. Using accelerated screening techniques, most weak units can be identified within a few days. This effort has been quite successful for manufacturers having good quality programs, and since the 1990s, the end user rarely sees most of these failures.

The middle portion of the bathtub has a failure rate that remains relatively flat or declining as a function of operating time interval. Failures are primarily due to random stresses in the environment. During this period of time it is reasonable to assume that the failure rate is constant. While many consider this to be too conservative due to the fact that the failure rate is probably declining, this assumption simplifies the math and is very appropriate for probabilistic SIF verification.

The end of the bathtub curve occurs when the strength of the product declines (susceptibility increases). This is commonly known as "wearout." Wearout occurs after several years but the mechanisms vary considerably depending on the type of component.

When the susceptibility of a component goes up, the failure rate increases rapidly. Manufacturers of instrumentation for safety instrumented applications will typically publish the useful life numbers for their products. A mechanical integrity program needs to address all instruments in safety applications before the end of their useful life.

EXAMPLE 3-2

Problem: Failure records at a plant indicate that a 28 watt DC solenoid that is constantly energized during normal operation of a process will almost always fail in three to five years. What needs to be added to a mechanical integrity program?

Solution: The mechanical integrity program of the process unit must include full replacement of all solenoids after three years of operation. NOTE: This is before the solenoid tends to fail due to wearout mechanisms.

Estimating Failure Rates

Often, time to failure data is not available for a collection of components. Incomplete data can be used to estimate failure rates under these circumstances but one must be very careful, especially when estimating failure rate data to be used for probabilistic SIF verification. Results that are not conservative can lead designers to believe that safety integrity levels are higher than they really are.

Consider the data from Table 3-1. If it is assumed that the first four failures would be detected by the manufacturer or would occur during the first two weeks of installation and commissioning, they can be excluded from a "useful life" failure rate calculation. If only the data from the first year of operation (8760 hours) is available for failure rate estimation, the failure data is presented in Table 3-2.

Table 3-2. Limited Time to Failure Data from Table 3-1

Units Operating	Failure Number	Hours
26	5	409
25	6	614
24	7	831
23	8	1045
22	9	1282
21	10	1540
20	11	1815
19	12	2106
18	13	2414
17	14	2740
16	15	3091
15	16	3471
14	17	3886
13	18	4348
12	19	4862
11	20	5431
10	21	6056
9	22	7499
8	23	8339

Given this data, a conservative way to estimate the constant failure rate is to calculate the average time to failure. It can be shown that a constant failure rate is related to MTTF (mean time to failure) according to the equation below (Ref 6, pg. 73).

$$\lambda = 1/\text{MTTF} \tag{3-3}$$

For the data set of Table 3-2, it can be seen that nineteen (19) failures are listed. Given that the data started with twenty-six (26) units operating, seven units are still operating successfully at the end of the year. The MTTF for those seven units operating successfully at one year is not known. If it is conservatively assumed that all seven will operate for at least one more hour, then an MTTF of 8761 could be used for those seven units.

Given those numbers, the average value for time to failure (MTTF) equals 6479 hours. The constant failure rate would be 0.000162 failures per hour. This is calculated by adding the actual time to failure for each of the nineteen failures, plus the number 8761 for the seven successful units, and dividing by nineteen, the number of failures. This is shown in Table 3-3.

Of course, we could calculate the actual operating hours from the data set of Table 3-2. The failure rate could then be obtained by dividing the

Table 3-3. MTTF Calculation

Hours
409
614
831
1045
1282
1540
1815
2106
2414
2740
3091
3471
3886
4348
4862
5431
6056
7499
8339
8761
8761
8761
8761
8761
8761
8761

Sum	123106
# Failures	19
MTTF	6479

number of failures by the operating hours. A detailed operational hour calculation is shown in Table 3-4.

Table 3-4. Operational Hours

Units Operating	Failure Number	Hours	Operating hours
26	5	409	10634
25	6	614	5125
24	7	831	5208
23	8	1045	4922
22	9	1282	5214
21	10	1540	5418
20	11	1815	5500
19	12	2106	5529
18	13	2414	5544
17	14	2740	5542
16	15	3091	5616
15	16	3471	5700
14	17	3886	5810
13	18	4348	6006
12	19	4862	6168
11	20	5431	6259
10	21	6056	6250
9	22	7499	12987
8	23	8339	6720
7		Total	120152

Nineteen failures occurred during this period, so the failure rate would be:

19 failures/120,152 hours = 0.000158 failures per hour

This number is close to the failure rate calculated above but does not make the assumption about the remaining seven units operating only one more hour and is, therefore, less conservative.

Censored Data

There are times where the only data available is the quantity of failures for a given quantity of units over a time period. For example, the data from Table 3-2 might be summarized. The summary report might say, "After installation and commissioning, twenty-six units were placed into operation. There were 19 failures for one year of operation." Some take this data and do an incorrect calculation. The operational hours are estimated by saying that 26 units run for one year – 26 × 8760 = 227,760 hours. The failure rate is, therefore:

19 failures/227,760 hours = 0.00008 failures per hour

The problem with this method is that it implicitly assumes that all units fail at the very end of the time period. This is terribly optimistic and results in an incorrect low estimate of the failure rate.

One could assume that all failures occur in the first hour of the time period. In that case, the operational hours could be estimated as follows:

$$(26 - 19) \text{ units} \times 8760 \text{ hours} = 61{,}320 \text{ hours}$$

The failure rate is, therefore:

$$19 \text{ failures} / 61{,}320 \text{ hours} = 0.00031 \text{ failures per hour}$$

Lastly, one could assume that all the failures occurred in the middle of the time period. The operating hours would be calculated as:

$$19 \times 8760/2 + 7 \times 8760 = 144{,}540 \text{ hours}$$

The failure rate is, therefore:

$$19 \text{ failures} / 144{,}540 \text{ hours} = 0.000131 \text{ failures per hour}$$

The mid-range value seems reasonable, but it is low when compared with the more precise numbers obtained with the detailed data. Clearly the most conservative choice is to assume all failures occur in the first hours when confronted with censored data. The authors recommend this approach.

Confidence Factor

Operating and maintenance data is often not of a sufficient quality to support the determination of failure rates. When it is available, it is necessary to have the quantities needed to have good statistical confidence in the calculated result. Therefore, functional safety standards like IEC 65108 (Ref. 1, Part 2, Annex C) require that a "single-sided ... confidence interval of at least 70%" be used when evaluating failure data. This requirement recognizes that one may not have enough data to achieve statistical accuracy. Given a limited quantity of failure rate data, the actual failure rate may be larger than the calculated result.

A confidence interval can be calculated that shows the probability that the actual failure rate is within the limits of the confidence interval. Figure 3-3 shows the situation for a single-sided 70% confidence interval.

Most reliability engineers use a Chi-Squared probability distribution for the uncertainty. The Chi-Squared distribution is a special case of the Gamma distribution and is appropriate for evaluating uncertainty (Ref. 7, Appendix 5, and Ref 8, Chapter 5). Given a point estimate of the failure

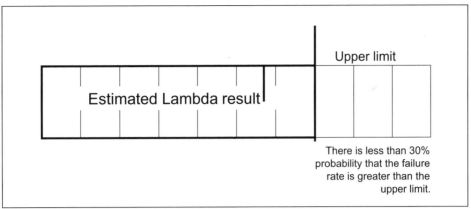

Figure 3-3. Confidence Interval of 70%

rate of 0.000158 from above, the upper limit failure rate can be calculated as follows:

$$\frac{1}{(2 \times 120,152) \times X^2(1 - 0.7, 2 \times (19 + 1))} = 0.000184 \text{ failures per hour}$$

where 120,152 equals the number of operational hours and 19 equals the number of failures.

The Chi-Squared function was used with the 70% confidence interval and 40($2 \times (19 + 1)$) degrees of freedom. The +1 is added to account for the possibility that another failure may occur in the next moment of operation.

Getting Failure Rate Data

It should be well understood that estimating failure rate data is complicated. In general for probabilistic SIF verification purposes, the less one knows, the more conservative one must be. If little data is available, then conservative choices must be made when choosing data. Fortunately, the quality of data is improving (see Chapter 8).

Exercises

3-1. A software bug causes a logic solver to fail in an unpredictable and apparently random manner. Will this failure be considered a random failure or systematic failure?

3-2. A system has a probability of failure (all modes) for each one-year mission time of 0.15. What is the probability of a failure for a ten-year mission time? (No wear out, etc.)

3-3. Unreliability for a system with one failure mode is given as 0.002. What is the reliability?

3-4. A module has an MTTF of 75 years for all failure modes. Assuming a constant failure rate, what is the total failure rate for all failure modes?

3-5. A module has an MTTF of 75 years. What is the reliability of this module for a time period of six months?

3-6. A transmitter has a total failure rate of 0.006 failures per year. What is the MTTF?

3-7. Which of the following likelihoods are best analyzed by statistical methods?

   I. The failure of a solenoid valve
   II. The explosion of a debutanizer distillation column
   III. The failure of an RTD temperature sensor
   IV. The rupture of a pressure vessel in normal service

   a. only I, II and IV
   b. only I and IV
   c. only I and II
   d. only I, III and IV

3-8. A manufacturer states that the failure rate for a pressure switch is constant and the MTBF is 2 years. Does this mean that:

   a. the expected minimum life for the switch is 2 years.
   b. the expected maximum life for the switch is 2 years.
   c. there is a 100% probability that the switch will survive for 2 years.
   d. there is a 50% probability that the switch will survive for 2 years.
   e. None of the above.

3-9. The total test time for 50 transmitters in identical service was 7640 hours. During this period, 16 failures were experienced. Calculate the failure rate.

3-10. The total test time for 50 transmitters in identical service was 7640 hours. During this period, 0 failures were experienced. Describe the procedure for calculating the failure rate.

REFERENCES AND BIBLIOGRAPHY

1. IEC 61508, *Functional Safety of electrical / electronic / programmable electronic safety-related systems*, Geneva: Switzerland, 2000.

2. ANSI/ISA-84.00.01-2004, *Functional Safety: Safety Instrumented Systems for the Process Industry Sector – Parts 1, 2, and 3 (IEC 61511 Mod)*, NC: Research Triangle Park, ISA, 2004.

3. Brombacher, A. C., *Reliability by Design*, John Wiley and Sons, UK: Chichester, 1992.

4. Beurden, I. J. W. R. J. van., *Stress-strength simulations for common cause modeling, Is physical separation a solution for common cause failures?*, PA: Spring House, Moore Products Co., 1997.

5. Goble, W. M., and Bukowski, J. V., "Verifying Common Cause Reduction Rules for Fault Tolerant Systems via Simulation using a Stress- Strength Failure Model," *ISA Transactions*, NC: Research Triangle Park, 2000.

6. Goble, W. M., *Control Systems Safety Evaluation and Reliability*, second edition, NC: Research Triangle Park: ISA, 1998

7. Billinton, Roy, and Allan, R. N., *Reliability Evaluation of Engineering Systems: Concepts and Techniques*, NY: New York, Plenum Press, 1983.

8. Henley, E. J., and Kumamoto, H., *Probabilistic Risk Assessment: Reliability Engineering, Design, and Analysis*, NJ: Piscataway, IEEE Press, 1992.

4
Basic Reliability Engineering

There are a number of common metrics used within the field of reliability engineering. Primary ones include reliability, unreliability, availability, unavailability, and MTTF. But, when different failure modes are considered — as they are when doing SIF verification — then new metrics are needed. These include PFS, PFD, PFDavg, MTTFS, and MTTFD.

In developing the applicable metrics concepts in this chapter, a single failure mode is assumed throughout. Multiple failure mode metrics are presented in a subsequent chapter.

Measurements of Successful Operation – No Repair

Probability of Success—This is often defined as the probability that a system will perform its intended function when needed and operated within its specified limits. The phrase at the end tells the user of the equipment that the published failure rates apply only when the system is not abused or otherwise operated outside its specified limits.

Using the rules of reliability engineering, one can calculate probability of successful operation for a particular set of circumstances. Depending on the circumstances, that probability is called "reliability" or "availability" (or on occasion, some other name).

Reliability—A measure of successful operation for a specified interval of time. Reliability, $R(t)$, is defined as "the probability that a system will perform its intended function when required to do so if operated within its specified limits for a specified operating time interval." The definition includes five important aspects.

1. The system's "intended function" must be known.
2. "When the system is required to function" must be judged.

3. "Satisfactory performance" must be determined.

4. The "specified design limits" must be known.

5. An operating time interval is specified.

Consider a newly manufactured and successfully tested component. It operates properly when put into service (T = 0). As the operating time interval (T) increases, it becomes less likely that the component will remain successful. Since the component will eventually fail, the probability of success for an infinite time interval is zero. Thus, all reliability functions start at a probability of one and decrease to a probability of zero (Figure 4-1).

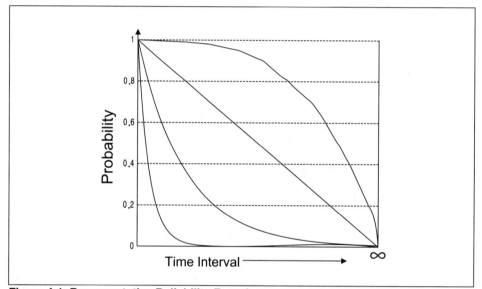

Figure 4-1. Representative Reliability Functions

Reliability is a function of the operating time interval. A statement such as "system reliability is 0.95" is meaningless because the time interval is not known. The statement, "The reliability equals 0.98 for a mission time of one hundred hours," makes perfect sense.

A reliability function can be derived directly from probability theory. Assume that the probability of successful operation for a one-hour time interval is 0.999. What is the probability of successful operation for a two-hour time interval? The system will be successful only if it is successful for both the first hour and the second hour. Therefore, the two-hour probability of success equals:

$$0.999 \times 0.999 = 0.998$$

The analysis can be continued for longer time intervals. For each time interval the probability can be calculated by the equation:

$$P(t) = 0.999^t$$

Figure 4-2 shows a plot of probability versus operating time per this equation. The plot is a reliability function.

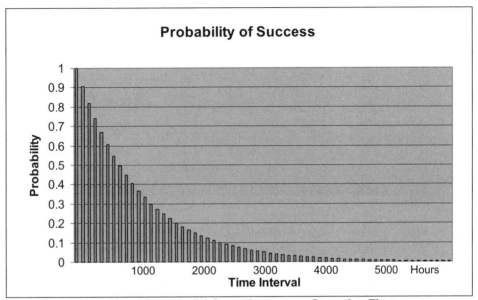

Figure 4-2. Probability of Successful Operation versus Operating Time

Reliability is a metric originally developed to determine probability of successful operation for a specific "mission time." For example: If a flight time is ten hours, a logical question is, "What is the probability of successful operation for the entire flight?" The answer would be the *reliability* for the 10 hours duration. It is generally a measurement applicable to situations where on-line repair is not possible, such as an unmanned space flight or an airborne aircraft.

When evaluating life test results, the reliability function is calculated by dividing the number of surviving units at any time by the total number of units starting the test.

$$R(t) = Ns/N \qquad (4\text{-}1)$$

where:
        Ns = number of surviving units
        N = number of total test units

Considering the test data from Table 3-1, the reliability function is shown in Figure 4-3.

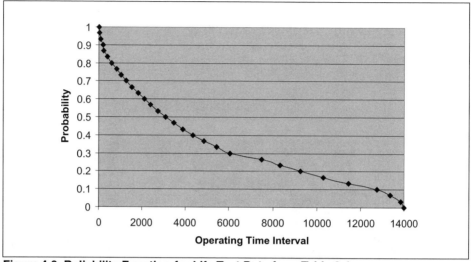

Figure 4-3. Reliability Function for Life Test Data from Table 3-1

Unreliability—A measure of failure, unreliability F(t) is defined as "the probability that a device will fail in the time interval from zero to T." Unreliability measures the probability that failure time will be less than or equal to the operating time interval. Since any component must be either successful or failed, F(t) is the one's complement of R(t).

$$F(t) = 1 - R(t) \qquad (4\text{-}2)$$

Figure 4-4 shows the unreliability function for the life test data from Table 3-1.

Mean Time To Failure (MTTF)—One of the most widely used reliability parameters is the MTTF. It has been formally defined as the "expected value" of the random variable Time To Fail, T. Unfortunately, the metric has evolved into a confusing number. MTTF has been misused and misunderstood. It has been misinterpreted as "guaranteed minimum life."

Formulas for MTTF are derived and often used for products during the useful life period. This method excludes wearout failures. This often results in a situation in which the MTTF is 300 years and useful life is only 40 years. Note that instruments should be replaced before the end of their useful life. When this is done, the mean time to random failures will be similar to the number predicted.

Ask an experienced plant engineer, "What is the MTTF of a pressure transmitter?" This engineer would likely include wearout and might answer, "Thirty-five years." Then the engineer would look at the specified

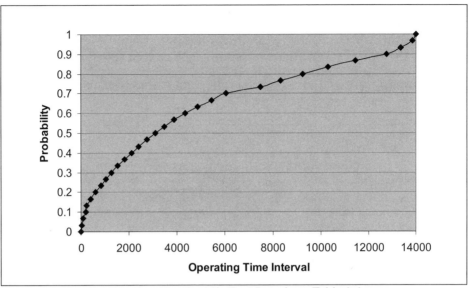

Figure 4-4. Unreliability Function for Life Test Data from Table 3-1

EXAMPLE 4-1

Problem: A component has a reliability of 0.999 for a one-hour mission time. What is the unreliability?

Solution: F = 1 − 0.999 = 0.001

MTTF of 300 years and think that the person who calculated that number should come out and stay with him for a few years and see the real world.

Generally, the term MTTF is being defined during the useful life of a device. "End of life" failures are generally not included in the number.

Constant Failure Rate—When a constant failure rate is assumed (which is valid during the useful life of a device), then the relationship between reliability, unreliability, and MTTF are straightforward. If the failure rate is constant, then:

$$\lambda(t) = \lambda \qquad (4\text{-}3)$$

For that assumption it can be shown that:

$$R(t) = e^{-\lambda t} \qquad (4\text{-}4)$$

$$F(t) = 1 - e^{-\lambda t} \qquad (4\text{-}5)$$

and

$$\text{MTTF} = 1/\lambda \qquad (4\text{-}6)$$

Figure 4-5 shows the reliability and unreliability functions for a constant failure rate of 0.001 failures per hour. Note that the plot for reliability looks the same as Figure 4-2, which shows the probability of successful operation given a probability of success for one hour of 0.999.

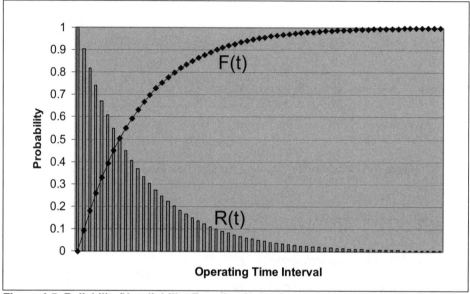

Figure 4-5. Reliability/Unreliability Function Using λ = 0.001 Failures per Hour

It can be shown that a constant probability of success is equivalent to an exponential probability of success distribution as a function of operating time interval.

EXAMPLE 4-2

Problem: A device has a constant failure rate of 5000 FITS during its useful life. What is the MTTF?

Solution: 5000 FITS equals 0.000005 failures per hour. The MTTF equals 1/0.000005 = 200,000 hours. In years this equals 200,000/8760 = 22.83.

> **EXAMPLE 4-3**
>
> **Problem**: A pressure transmitter has an MTTF of 250 years. What is the failure rate in failures per year and FITS?
>
> **Solution**: The failure rate per year equals 1/MTTF = 1/250 = 0.004 failures per year. To convert to FITS which equal 10^{-9} failures per hour, 0.004/8760 hours per year = 4.57×10^{-7} = 457 FITS.

> **EXAMPLE 4-4**
>
> **Problem**: A pressure transmitter has an MTTF of 250 years. What is the reliability for a mission time of 5 years?
>
> **Solution**: The reliability R(t) equals exp(-(1/250) × 5) = 0.98

Useful Approximations

Mathematically it can be shown that certain functions can be approximated by a series of other functions. For all values of x, it can be shown that

$$e^x = 1 + x + x^2/2! + x^3/3! + x^4/4! + \ldots \tag{4-7}$$

For a sufficiently small value of x, the exponential can be approximated with

$$e^x = 1 + x$$

Substituting $-\lambda t$ for x,

$$e^{\lambda t} = 1 + \lambda t \tag{4-8}$$

Thus, there is an approximation for unreliability when λt is sufficiently small:

$$F(t) = \lambda t \tag{4-9}$$

Remember that this is only an approximation and not a fundamental equation. Often the notation for unreliability is PF (probability of failure) and the equation is shown as

$$PF(t) = \lambda t \tag{4-10}$$

> **EXAMPLE 4-5**
>
> **Problem**: A device has a failure rate of 0.00005 failures per hour. For a mission time of 100 hours, what is the unreliability?
>
> **Solution**: Using the exponential equation, the unreliability equals $1 - \exp(-0.00005 \times 100) = 0.00498$. Using the approximation, the unreliability equals $0.00005 \times 100 = 0.005$. Note that the approximation always results in a pessimistic answer.

Measurements of Successful Operation—Repairable Systems

The measurement "reliability" requires that a system be successful for an interval of time. While this probability is a valuable estimate for situations in which a system cannot be repaired during a mission, something different is needed for an industrial process control system where repairs can be made, often with the process operating.

Mean Time To Restore (MTTR)—MTTR is the "expected value" of the random variable "restore time" (or time to repair). The definition includes the time required to detect that a failure has occurred as well as the time required to make a repair once the failure has been detected and identified. Like MTTF, MTTR is an average value. MTTR is the average time required to move from unsuccessful operation to successful operation.

In the past, the acronym MTTR stood for "Mean Time To Repair." The term was changed in IEC 61508 because of confusion as to what was included. Some thought that Mean Time To Repair included only actual repair time. Others interpreted the term to include both time to detect a failure (diagnostic time) and actual repair time. The term "Mean Dead Time (MDT)" is commonly used in some parts of the world and means the same as Mean Time To Restore.

Mean Time To Restore (MTTR) is a term created to clearly include both diagnostic detection time and actual repair time. Of course, when actually estimating MTTR one must include time to detect, recognize, and identify the failure; time to obtain spare parts; time for repair team personnel to respond; actual time to do the repair; time to document all activities, and time to get the equipment back in operation.

Reliability engineers often make the assumption that the probability of repair is an exponentially distributed function in which case the "restore rate" is a constant. The lower case Greek letter mu is used to represent restore rate by convention. The equation for restore rate is:

$$\mu = 1/\text{MTTR} \tag{4-11}$$

Restore times can be difficult to estimate. This is especially true when periodic activities are involved. Imagine the situation in which a failure in the safety instrumented system is not noticed until a periodic inspection and test are done. The failure may occur right before the inspection and test, in which case the detection time might be near zero. Or it may occur right after the inspection and test, in which case the detection time may get as large as the inspection period.

In such cases it is probably best to model repair probability as a periodic function rather than as a constant (Ref. 1). This is discussed in the following section, "Average Unavailability with Periodic Inspection and Test."

EXAMPLE 4-6

Problem: The plant maintenance team has made the following estimates:

> Average time to detect that a failure has occurred in the basic process control system = 18 hours.
>
> Average time to obtain spare parts = 24 hours.
>
> Average time to make the repair and test the system = 8 hours.

What is the Mean Time To Restore? What is the restore rate?

Solution: Assuming these are sequential operations, the MTTR is the sum of the estimated times. MTTR equals 50 hours. The restore rate is 0.02 restores per hour.

Mean Time Between Failures (MTBF)—MTBF is defined as the average time period of a failure/repair cycle. It includes time to failure, any time required to detect the failure, and actual repair time. This implies that a component has failed and then has been successfully repaired. For a simple repairable component,

$$MTBF = MTTF + MTTR \qquad (4\text{-}12)$$

The MTBF term can also be confusing. Since MTTR is usually much smaller than MTTF, MTBF is approximately equal to MTTF. The term MTBF is often substituted for MTTF and applies to both repairable systems and non-repairable systems.

Availability—The reliability measurement was not sufficiently useful for engineers who needed to know the average chance of success of a system when repairs are possible. Another measure of system success for repairable systems was needed. That metric is "Availability." Availability is defined as "the probability that a device is successful at time t when needed and operated within specified limits." No operating time interval

> **EXAMPLE 4-7**
>
> **Problem**: An industrial I/O module has an MTTF of 87,600 hours. When the module fails, it takes an average of 2 hours to repair. What is the MTBF?
>
> **Solution**: Using equation (4-11), the MTBF = 87,602 hours. The MTBF is approximately equal to the MTTF.

is directly involved. If a system is operating successfully, it is available. It does not matter whether it has failed in the past and has been repaired or has been operating continuously from startup without any failures. Availability is a measure of "uptime" in a system, unit, or module.

Availability and reliability are different metrics. Reliability is always a function of failure rates and operating time interval. Availability is a function of failure rates and repair rates. While instantaneous availability will vary during the operating time interval, this is due to changes in failure probabilities and repair situations. Availability is often calculated as an average over a long operating time interval. This is referred to as "steady state availability."

In some systems, especially safety instrumented systems, the repair situation is not constant. In safety instrumented systems the situation occurs when failures are discovered and repaired during periodic inspection and test. For these systems, steady state availability is NOT a good measure of system success. Instead, average availability is calculated for the operating time interval between inspections. NOTE – This is not the same measurement as steady state availability.

Unavailability—a measure of failure that is used primarily for repairable systems. It is defined as "the probability that a device is not successful (is failed) at time t." Different metrics can be calculated, including steady state unavailability and average unavailability over an operating time interval. Unavailability is the one's complement of availability; therefore,

$$U(t) = 1 - A(t) \qquad (4\text{-}13)$$

> **EXAMPLE 4-8**
>
> **Problem**: A controller has a steady state availability of 0.99. What is the steady state unavailability?
>
> **Solution**: Using equation (4-13), Unavailability = 1 – 0.99 = 0.01.

Steady State Availability—Traditionally, reliability engineers have assumed a constant repair rate. When this is done, probability models can be solved for "steady state" or average probability of successful operation. The metric can be useful, but it has relevance only for long intervals of time.

Figure 4-6 shows a Markov probability model of a single component with a single failure mode. This model can be solved for steady state availability and steady state unavailability.

$$A = MTTF/(MTTF + MTTR) \qquad (4\text{-}14)$$

$$U = MTTR/(MTTF + MTTR) \qquad (4\text{-}15)$$

When the Markov model of Figure 4-6 is solved for availability as a function of operating time interval, the result is shown in Figure 4-7, labeled A(t). It can be seen that the availability reaches a "steady state" after some period of time.

Figure 4-8 shows a plot of unavailability versus unreliability.

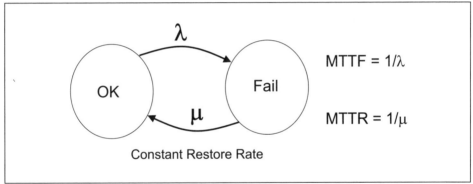

Figure 4-6. Markov Model for Single Component and Single Failure Mode

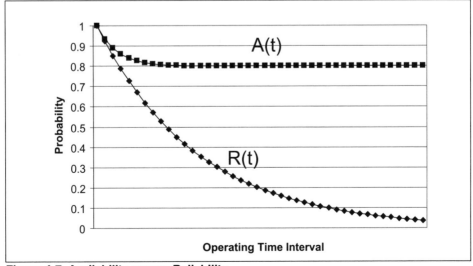

Figure 4-7. Availability versus Reliability

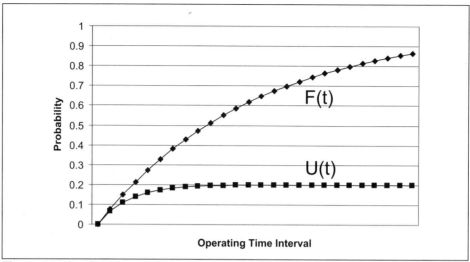

Figure 4-8. Unreliability vs. Unavailability

Average Unavailability with Periodic Inspection and Test

In low demand safety instrumented system applications, the restore rate is NOT constant. For failures not detected until a periodic inspection and test, the restore rate is zero until the time of the test. If it is discovered that the system is operating successfully, then the probability of failure is set to zero. If it is discovered that the system has a failure, it is repaired. In both cases, the restore rate is high for a brief period of time. Dr. Julia V. Bukowski has described this situation and proposed modeling repair as a periodic impulse function (Ref. 1). This method is described in Appendix G.

Figure 4-9 shows a plot of probability of failure in this situation. This can be compared with unavailability calculated with the constant restore rate model as a function of operating time. With the constant restore model, the unavailability reaches a steady state value. This value is clearly different from the result that would be obtained by averaging the unavailability calculated using a periodic restore period.

It is often assumed that the periodic inspection and test will detect all failed components and the system will be renewed to perfect condition. Therefore, the unreliability function is suitable for the problem. A mission time equal to the time between periodic inspection and test is used. In safety instrumented system applications, the objective is to find a model for the probability that a system will fail when a dangerous condition occurs. This dangerous condition is called a "demand."

Our objective then is to calculate the probability of failure on demand. If the system is operating in an environment where demands are infrequent (for example once per ten years) and independent from system proof tests, then an average of the unreliability function will provide the average

Basic Reliability Engineering   55

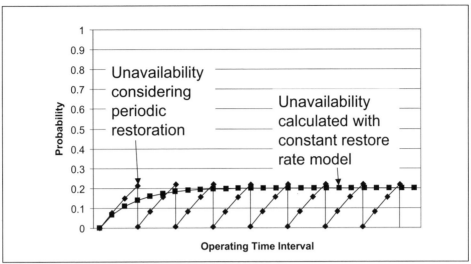

Figure 4-9. Probability of Failure with Periodic Testing

probability of failure. This by definition is an unavailability function since repair is allowed. (NOTE – This averaging technique is not valid when demands are more frequent. Special modeling techniques are needed in that case.)

As an example, consider the single component unreliability function given in equation (4-5).

$$F(t) = 1 - e^{-\lambda t}$$

This can be approximated as explained previously with equation (4-8).

$$F(t) = \lambda t$$

The average can be obtained by using the expected value equation

$$PFavg = \frac{1}{T}\int_0^T PF(t)dt \qquad (4\text{-}16)$$

with the result being an approximation equation

$$PFavg = \lambda t/2 \qquad (4\text{-}17)$$

For a single component (non-redundant) or a single channel system, the approximation is shown in Figure 4-10. Note that the approximation is conservative and supplies a pessimistic value.

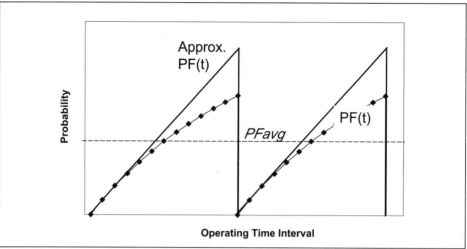

Figure 4-10. Average Probability of Failure Approximation with Periodic Testing

---

EXAMPLE 4-9

Problem: A transmitter has a failure rate of 0.005 failures per year. What is the average probability of failure if the transmitter is 100% tested and calibrated every two years?

Solution: Using equation (4-16), if one assumes perfect inspection and testing, then average probability of failure = 0.005 × 2/2 = 0.005.

Periodic Restoration and Imperfect Testing

It is quite unrealistic to assume that inspection and testing processes will detect all failures. In the worst case, assume that testing is not done. In that situation what is the mission time? If the equipment is used for the life of an industrial facility, plant life is the mission time. Probability of failure would be modeled with the unreliability function using the plant life as the time interval.

If the equipment is required to operate only on demand and the demand is independent of system failure, the unreliability function can be averaged as explained in the preceding section.

When only some failures are detected during the periodic inspection and test, the average probability of failure can be calculated using an equation that combines the two types of failures: those detected by the test and those undetected by the test. One must estimate the percentage of failures detected by the test to make this split. The equation would be:

$$PFavg = C_{PT} \lambda \, TI/2 + (1 - C_{PT}) \lambda \, LT/2 \qquad (4\text{-}18)$$

Basic Reliability Engineering

> **EXAMPLE 4-10**
>
> **Problem**: A valve used in a safety instrumented system has a failure rate of 0.002 failures per year. The valve is not tested. It will be used during the lifetime of the process unit, 25 years. The demand is estimated to be once every 100 years. What is the probability of failure during a demand?
>
> **Solution**: Equation (4-16) gives the average probability of failure. This is appropriate for this problem using a time interval of 25 years. The average probability of failure = 0.002 × 25 / 2 = 0.025.

where:

λ = the failure rate
C_{PT} = the percentage of failures detected by the proof test
TI = the periodic test interval
LT = lifetime of the process unit

> **EXAMPLE 4-11**
>
> **Problem**: A valve is used in a safety instrumented system. It has a failure rate of 0.002 failures per year. The valve is tested each year. The inspection and test will detect broken valve stems, tight packing, jammed actuators and leakage. Seat leakage and sealing ability of the valve are not tested. It is estimated that 70% of the failures are detected during that test. The valve is used for the lifetime of the process unit, 25 years. The demand is estimated to be once every 100 years. What is the average probability of failure?
>
> **Solution**: Equation (4-17) is used. The average probability of failure equals
>
> (0.7) × 0.002 × 1/2 + (1 − 0.7) × 0.002 × 25/2 = 0.0082

Exercises

4-1. An instrument has a failure rate of 0.015 failures per year. What is the unreliability for a five-year mission?

4-2. An instrument has a failure rate of 0.015 failures per year. All failures are immediately detectable. The repair time average is 24 hours. What is the steady state unavailability?

4-3. An instrument has a failure rate of 0.015 failures per year. Failures are detected only when a periodic inspection is done once per year. What is the PFavg?

4-4. A valve has a failure rate of 0.06 failures per year. A periodic inspection done once a year can detect 60% of the failures. The valve is operated for ten years before it is removed from service and overhauled. What is the PFavg for the ten-year operational interval?

4-5. During an annual inspection of a safety instrumented system, 5 of the 112 safety functions had dangerous failures on similar solenoid valves. What number should be assigned to the probability of dangerous failure for this collection of solenoid valves?

   a. 0.05
   b. 0.045
   c. 0.043
   d. 0.0112
   e. 0.1

4-6. A system has a transmitter, a controller, and a valve. The probability of failure for the next five years equals 0.15 for the transmitter, 0.008 for the controller, and 0.19 for the valve. For the next five-year time interval, what is the probability of system success?

   a. 0.683
   b. 0.348
   c. 0.652
   d. 0.352
   e. 0.545

4-7. An instrument has a MTTF of 28,000 hours and a MTTR of 48 hours. What is the MTBF?

4-8. A single pressure transmitter is used to initiate a trip in a safety instrumented function. It has a dangerous failure rate of 5×10^{-6} failures per hour. None of the failures are detected by any automatic diagnostics. It is inspected every two years, and all failures are detected during this inspection. If PFDavg is the PFavg counting only dangerous failures, what is the PFDavg?

4-9. A system is repairable. Restore time averages 48 hours, as all failures are immediately recognizable. The percentage of downtime must be calculated for a 20-year expected system life. What is the best measure to use from reliability engineering?

4-10. A safety instrumented system can have failures that are not immediately known. Therefore, it is inspected every two years. Which metric is the best measure of probability of failure?

REFERENCES AND BIBLIOGRAPHY

1. Bukowski, J. V., "Modeling and Analyzing the Effects of Periodic Inspection on the Performance of Safety-Critical Systems," *IEEE Transactions of Reliability*, Vol. 50, No. 3, NY: New York, IEEE, September 2001.

5
System Reliability Engineering

Introduction

Given the failure rate of a component and an operational time interval (mission time), one can calculate the reliability of that component. If the component is repairable and the restore rate is estimated, the steady state availability of the component can be calculated. If the failure rates, proof test interval, proof test coverage, and component lifetime are known, one can calculate the average probability of failure.

But the problem is usually much more complicated. Real problems involve systems of many components. Many of these systems have "redundancy" designed into the system in order to maintain successful system operation even when a component fails. How does one calculate reliability or availability of these complex systems?

Several different modeling techniques have been developed. In this chapter block diagrams (also called network modeling), fault trees, and Markov models are presented in a simple introductory way. More advanced and realistic modeling techniques are covered in later chapters.

System Model Building

A systematic approach to model building should be used regardless of the modeling technique (Ref. 1, Chapter 13, page 307). The following steps should be included in one form or another.

1. Define "what is the intended function?" One must effectively define what components are included in the model. This is normally done by defining failure and excluding components not needed to achieve the desired function.

2. Obtain failure rate and failure mode data for each component in the system and create a checklist of all components with their failure modes.

3. Understand how the system works and how the failure of each component in each mode will affect the system. A system failure modes and effects analysis (FMEA) is often done to accomplish this.

4. Build the model using the checklist from above to ensure that all components and all relevant failure modes are included in the model.

Reliability Block Diagrams

Many reliability engineers use block diagrams to represent the construction of a system from a reliability perspective. Although these diagrams are often called Network Models (Ref. 1, 2, and 3), other documents (Ref. 4 and 5) including the ISA technical report TR84.00.02 call the modeling technique a Reliability Block Diagram.

A Reliability Block Diagram (RBD) is a method of graphically showing probability combinations. A drawing is done with boxes representing the components of a system. An RBD is quite effective for two-state reliability problems. "Two-state" means that the system has two operational states, success and failure.

Although the two-state method could be used for multiple state problems such as safety instrumented systems, most find the technique confusing and error prone for such problems. The RBD technique is presented briefly here in order to build understanding of system reliability modeling.

The overall concept in the RBD method is that a system is successful if one can find a path across the drawing as indicated in Figure 5-1. If no path is available, the system has failed. Each block represents a component in the system. If the component is operating, one can pass through the block. If the component has failed, passage is not allowed. There are two fundamental constructs: series and parallel.

Series System

The most basic construct in a RBD is a series system (Figure 5-2). In this system all components are needed for the system to be successful. If any component fails, the system has failed.

The reliability of the system equals the product of the reliability of the components, since the system is successful only if all components are successful.

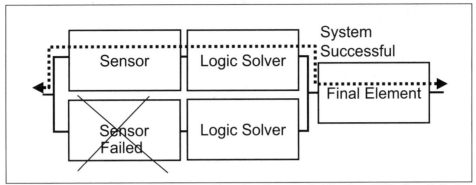

Figure 5-1. Reliability Block Diagram Showing Successful System With One Failed Component

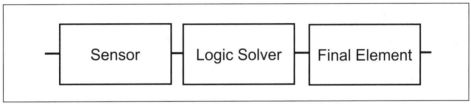

Figure 5-2. Reliability Block Diagram—Series System

$$Rs = Ra \times Rb \times Rc \times \ldots \qquad (5\text{-}1)$$

The unreliability of the system is given by the "union" of component unreliability. For a two-component system, the equation is:

$$Fs = Fa + Fb - Fa \times Fb \qquad (5\text{-}2)$$

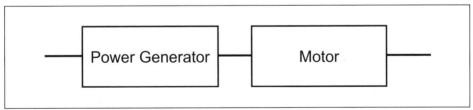

Figure 5-3. Reliability Block Diagram for Example Problem 5-1

Parallel System

The second basic type of construct in a RBD is the parallel network. The parallel drawing (Figure 5-4) shows that a system will be successful if any of the components are successful.

The reliability for a parallel system is the "union" of the reliabilities for the components. This is because a two-component system will be reliable if

EXAMPLE 5-1

Problem: A system consists of a power generator and a motor (Figure 5-3). The system is successful only if both components are successful. The power generator has an availability of 0.95 and the motor has an availability of 0.9. What is the system availability?

Solution: Equation 5-1 shows that system reliability equals the product of component reliabilities. The problem provides availabilities, not reliabilities, but the same equation applies as the RBD method is a probability combination method. Therefore, the system availability equals

$$A_s = 0.95 \times 0.9 = 0.855$$

EXAMPLE 5-2

Problem: A system consists of three sensors, a controller, a solenoid and an air-operated valve. The system is successful only if all components are successful. The sensors are identical and have steady state availabilities of 0.96. The controller has a steady state availability of 0.999. The solenoid has a steady state availability of 0.9 during its useful life of 5 years. The valve has a steady state availability of 0.9. What is the system steady state availability?

Solution: The system availability is the product of the component availabilities. In this case

$$A_s = 0.96 \times 0.96 \times 0.96 \times 0.999 \times 0.9 \times 0.9 = 0.716$$

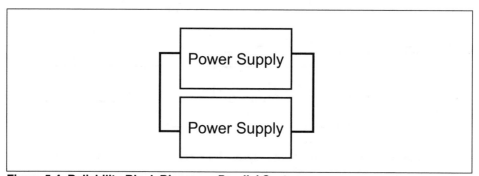

Figure 5-4. Reliability Block Diagram—Parallel System

either component A *or* component B is reliable. For a two-component system, the equation is:

$$R_s = R_a + R_b - (R_a \times R_b) \qquad (5\text{-}3)$$

The unreliability of the system equals the product of the unreliability of the components, since the system will fail only if all components fail.

$$Fs = Fa \times Fb \times Fc \times \ldots \tag{5-4}$$

> **EXAMPLE 5-3**
>
> **Problem**: A system consists of two power supplies (Figure 5-4). The system is successful only if either component is successful. The two power supplies are the same and have a reliability of 0.95 for a one-year mission time. What is the system reliability?
>
> **Solution**: Equation 5-3 shows that system reliability equals the union of component reliabilities. Therefore, the system reliability equals
>
> $$Rs = 0.95 + 0.95 - (0.95 \times 0.95) = 0.9975$$

> **EXAMPLE 5-4**
>
> **Problem**: A system consists of two power generators and two motors (Figure 5-5). The system is successful if one set of components is successful. The power generator has an availability of 0.95 and the motor has an availability of 0.9. What is the system availability?
>
> **Solution**: Using Equation 5-1 to obtain the reliability of each series path in combination with Equation 5-3 gives the solution
>
> $$Rs = (0.95 \times 0.9) + (0.95 \times 0.9) - [(0.95 \times 0.9) \times (0.95 \times 0.9)] = 0.9789$$

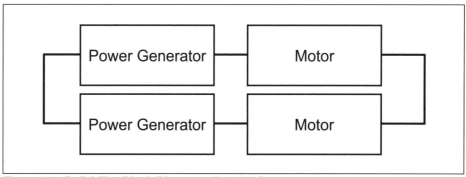

Figure 5-5. Reliability Block Diagram—Parallel System

Fault Trees

Another common technique for showing probability combinations is called a fault tree. This technique begins with the definition of an "undesirable event," usually a system failure of some type. The analyst continues by identifying all events and combinations of events that result in the identified undesirable event. The fault tree is therefore quite useful when modeling failures in a specific failure mode. These different failure modes can be identified as different undesirable events in different fault trees.

The fault tree method requires that one define an undesirable event (often called the "top event"). Consider the equipment set used for the safety instrumented function in Figure 5-6. A fault tree drawing shown in Figure 5-7 shows a top event defined as probability of failure on demand for the safety instrumented function shown in Figure 5-6.

A separate fault tree can be created for each failure mode. This is useful in safety instrumented function verification. Figure 5-8 shows a fault tree for a spurious trip of the same safety instrumented function from Figure 5-6. Note that these two fault trees are models of two different failure modes in the same safety instrumented function.

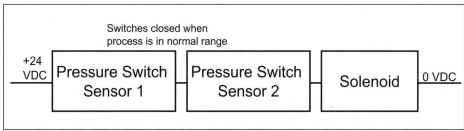

Figure 5-6. Safety Instrumented Function Equipment Set

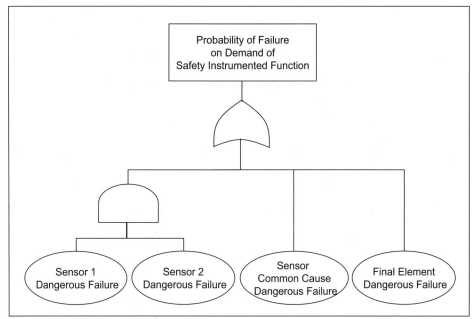

Figure 5-7. PFD Fault Tree for a SIF

System Reliability Engineering   67

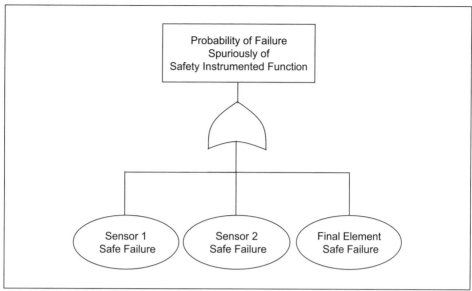

Figure 5-8. PFS Fault Tree for a SIF

Fault Tree Symbols

Fault trees show the logical relationship of probabilities with a few basic symbols. The most commonly used symbols include the AND gate, the OR gate, the BASIC FAULT (a fault that cannot be decomposed further), and the RESULTING FAULT. These are shown in Figure 5-9. There are additional symbols, some of which are also shown in Figure 5-9. IEC 61025 (Ref. 6), Fault Tree Analysis, shows 15 symbols. Other texts (Ref. 7) show even more fault tree symbols. For SIF verification analysis, the common symbols are sufficient.

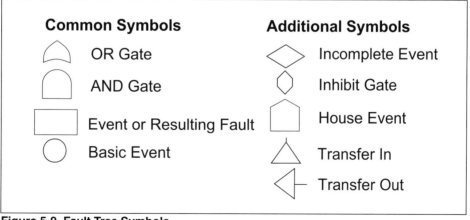

Figure 5-9. Fault Tree Symbols

Comparison of the Reliability Block Diagram and the Fault Tree

Both Reliability Block Diagrams (RBD) and Fault Trees are methods of graphically showing probability combinations. The primary difference is that the RBD modeler is focused on system success and the fault tree modeler is focused on a failure event. Figure 5-10 shows a fault tree AND gate and the equivalent RBD. Figure 5-11 shows a fault tree OR gate and the equivalent RBD. Note that the drawings clearly show the specific failure mode under consideration; therefore, the authors strongly prefer fault trees over reliability block diagrams when doing safety instrumented function verification calculations.

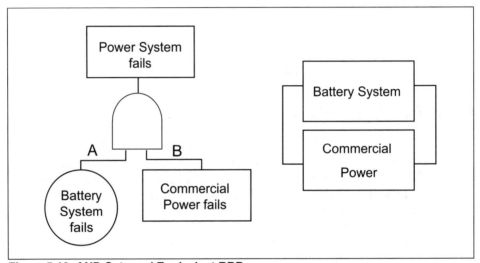

Figure 5-10. AND Gate and Equivalent RBD

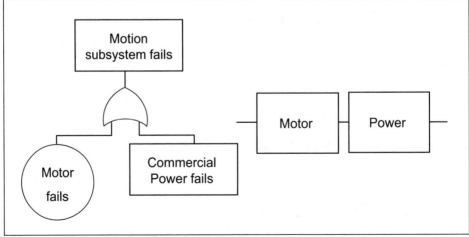

Figure 5-11. OR Gate and Equivalent RBD

Fault Tree AND Gates

Fault trees can be solved quantitatively like any probability combination method (Ref. 8). For a fault tree AND gate, all inputs must be present for the gate to be true or "active." For example, in Figure 5-10 with a two-input AND gate, both events A and B must occur for the gate be active. Referring to failure events, both failures must occur for the gate to be failed. If these events are independent (see Appendix B for a review of probability theory including definitions of independent events and mutually exclusive events), then the probability of getting failure event A and failure event B is given by the formula:

$$Fs = Fa \times Fb \tag{5-5}$$

Note that this is identical in format to Equation 5-4 showing the relationship from Figure 5-10.

EXAMPLE 5-5

Problem: A power system has a fault tree shown in Figure 5-10. Steady state unavailability of the battery system is estimated to be 0.01 (probability of failure at any moment in time). Steady state unavailability of commercial power is estimated to be 0.0001. What is the steady state unavailability of the power system?

Solution: The fault tree can be applied quantitatively. Since both power sources must fail for the system to fail, Equation 5-5 can be used.

Steady state unavailability for the system = $0.01 \times 0.0001 = 0.000001$

Fault Tree OR Gates

With an OR gate, when any of the inputs are true, then the gate output will be true or, as we have indicated before, active. Quantitative evaluation requires summation of the probabilities. For non-mutually exclusive events (the correct case for independent failures):

$$Fs = Fa + Fb - (Fa \times Fb) \tag{5-6}$$

Note that this is similar to Equation 5-2 again, showing the relationship between the fault tree and the equivalent RBD.

Approximation Techniques

Often, in order to speed up and simplify the calculation, the faults and events in a fault tree are sometimes assumed to be mutually exclusive and independent. Under this assumption, probabilities for the OR gates are

EXAMPLE 5-6

Problem: A motion subsystem consists of a motor and a power source. If steady state unavailability of a motor is 0.01 and steady state unavailability of the power source is 0.001, what is the steady state unavailability of the subsystem?

Solution: Equation 5-6 applies:

F(subsystem failure) = 0.01 + 0.001 − (.00001) = 0.01099

EXAMPLE 5-7

Problem: Three thermocouples are used to sense temperature in a reactor. The three signals are wired into a safety PLC, and a trip will occur if only one of the sensors indicates a trip. The probability of failure in the safe mode (causing a spurious trip) for a one-year mission time is 0.005. What is the probability of a spurious (false) trip?

Solution: An expanded version of Equation 5-6 is needed (see Appendix B).

F(subsystem failure) = Fa + Fb + Fc − (Fa × Fb) − (Fa × Fc) − (Fb × Fc) + (Fa × Fb × Fc)

F(subsystem failure) = 0.005 + 0.005 + 0.005 − (0.005 × 0.005) − (0.005 × 0.005) − (0.005 × 0.005) + (0.005 × 0.005 × 0.005) = 0.014925

added. Probabilities for the AND gates are multiplied. This approximation technique can provide rough answers when probabilities are low, as is the case when doing SIF verification. The approach is conservative when working with failure probabilities because the method gives an answer that is larger than the accurate answer.

EXAMPLE 5-8

Problem: Three thermocouples are used to sense temperature in a reactor as in Example 5-7. Use an approximation technique to estimate the probability of subsystem failure.

Solution: If mutually exclusive events are assumed, then probabilities can be added.

F(subsystem failure) = Fa + Fb + Fc = 0.005 + 0.005 + 0.005 = 0.015

This answer is more conservative (pessimistic) than the answer of Example 5-7, which was 0.014925.

Common Mistake

When building a fault tree, it is important to model individual components only once. It is easy to add something in twice, especially in complex systems. Consider the motion system fault tree of Figure 5-11. If two of these systems were required, the "combined" fault tree might be drawn incorrectly if the "commercial power" came from the same source. This is shown in Figure 5-12.

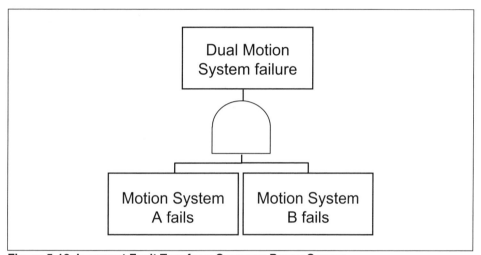

Figure 5-12. Incorrect Fault Tree for a Common Power Source

One might solve this fault tree using the answer from Example 5-6 as input. That fault tree provided a result of steady state unavailability = 0.01099 for the "motion sub-system." If that result were input into the fault tree of Figure 5-11, the result would be incorrectly approximated as steady state unavailability of the utility system of 0.000121.

If the "commercial power" came from the same source, the model should be drawn as shown in Figure 5-13. The result for this model is 0.0011, an order of magnitude higher, not 0.00012. Many fault tree software tools will identify potential problems of this type and if the commercial power was identified as one failure event, many tools would solve the problem correctly even if the tree was drawn incorrectly.

The problem with the fault tree of Figure 5-15 is that it shows the main valve failure multiple times. The fault tree could be drawn correctly to show the main valve failure only once (Figure 5-16).

This fault tree can be solved approximately using expanded versions of Equation 5-5 and Equation 5-6.

$$\text{PF (valve subsystem, one year)} = 0.0263 \times (0.0175 + 0.0175 + 0.0175 + 0.0175)$$
$$= 0.00184$$

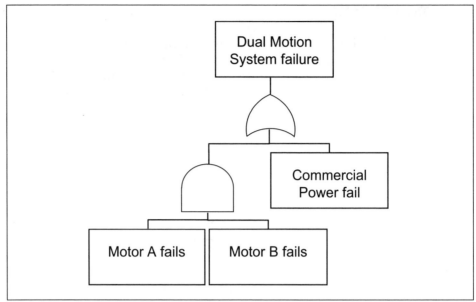

Figure 5-13. Correct Fault Tree for a Common Power Source

EXAMPLE 5-9

Problem: Consider a system with one main valve and four secondary valves (Figure 5-14). The system is successful if the main valve is successful or if all four of the secondary valves are successful. The system will fail if the main valve and one of the secondary valves fail. The main valve has one failure mode and a failure rate of 0.000003 failures per hour. The secondary valves have one failure mode and a failure rate of 0.000002 failures per hour. What is the probability of system failure if operated for a one-year mission time?

Solution: The probability of failure for defined interval of time is unreliability. The unreliability can be approximated using Equation 4-9.

For the main valve:

PF(one year) = 0.000003 failures per hour × 8760 hours per year = 0.0263

For each secondary valve:

PF(one year) = 0.000002 failures per hour × 8760 hours per year = 0.0175

A fault tree can be drawn to show probability combinations. Since the failure occurs if the main valve AND one secondary valve fail, the fault tree could be drawn as shown in Figure 5-14.

This fault tree can be solved approximately using a combination of Equations 5-5 and 5-6.

PF (valve subsystem, one year) = (0.0263 × 0.0175) + (0.0263 × 0.0175) + (0.0263 × 0.0175) + (0.0263 × 0.0175) = 0.00184

System Reliability Engineering   73

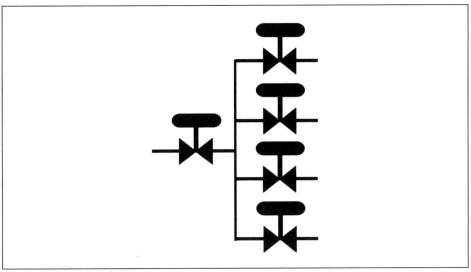

Figure 5-14. Valve Subsystem

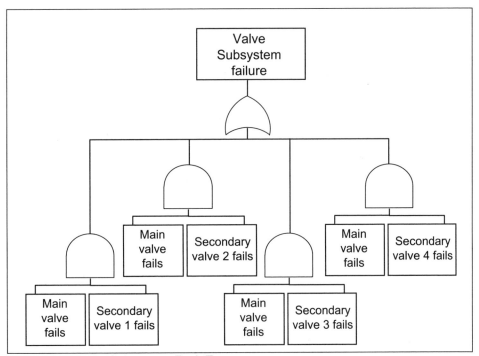

Figure 5-15. Valve Subsystem Fault Tree

This is the exact same answer as obtained in the fault tree of Figure 5-15 even though this fault tree has the main valve failure entered multiple times. In this case it can be seen that the math will produce the same results. Even so, one should draw fault trees with one failure event shown once on the drawing. That way there is no doubt about the interpretation of the model.

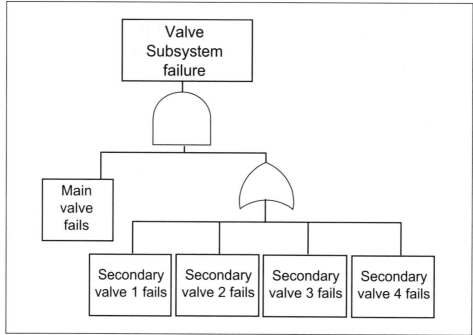

Figure 5-16. Alternative Valve Subsystem Fault Tree

Markov Models

Markov models are a reliability and safety modeling technique that uses state diagrams. These diagrams have only two simple symbols (see Figure 5-17): a circle representing a working or a failed system state and a transition arc representing a movement between states caused by a failure or a repair. Solution techniques for Markov models can directly calculate many different metrics compared to other reliability and safety evaluation techniques (Ref. 9).

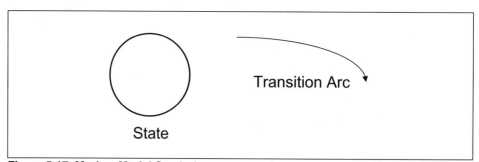

Figure 5-17. Markov Model Symbols

A Markov model example is shown in Figure 5-18. This model shows how the symbols are used. Circles (states) show combinations of successfully operating components and failed components. Possible component failures and repairs are shown with transition arcs, arrows that go from one state to another. A number of different combinations of failed and successful components are possible. Some represent system success states, while others represent system failure states. It should be noted that multiple failure modes can be shown on one drawing.

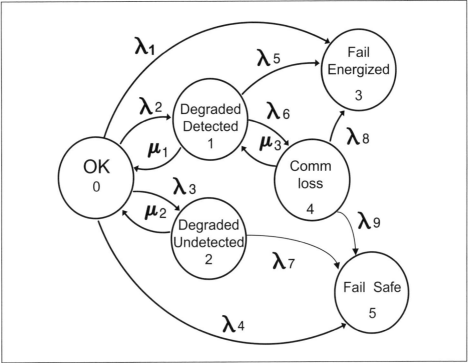

Figure 5-18. A Markov Model Example

Markov models are generally considered more flexible than other methods. On a single drawing, a Markov model can show the entire operation of a fault tolerant control system including multiple failure modes. Different repair rates can be modeled for different failure situations. If the model is created completely, it will show full system success states. It will also show degraded states where the system is still operating successfully but is vulnerable to further failures. The modeling technique provides clear ways to express failure sequences and can be used to model time dependent probabilities.

Markov Solution Techniques

A number of methods have been developed to solve Markov models. Some solution methods are not suitable for safety instrumented function verification. When periodic inspection and repair are performed, solving for steady-state unavailability is not correct. Numerical averaging of a discrete time model does work well, however.

One method well suited for spreadsheet numerical solution is discrete time matrix multiplication (Ref. 1, Chapter 8). A "time interval" is chosen, and failure rates are scaled for that time interval. The probability of a transition on each arc is given by the failure rate multiplied by the time interval. By choosing very small time intervals, the model can be solved with great accuracy. By convention, the scaled failure rate is listed above every transition arc instead of the probability, which includes failure rate and time interval. The time interval (Δt) is implied but rarely shown.

A Markov model can be represented by a square matrix showing the transition probabilities between each state. This transition matrix can be multiplied by a row matrix representing the starting state. The resulting row matrix gives time dependent state probabilities for each state. Consider a Markov model for the valve subsystem of Figure 5-14. A Markov model showing states and transitions is shown in Figure 5-19.

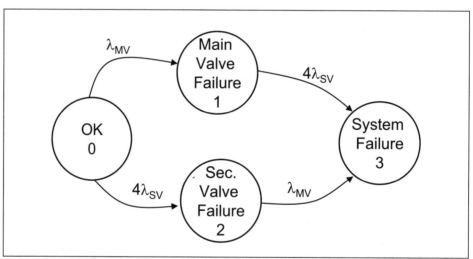

Figure 5-19. Markov Model of the Figure 5-14 Valve Subsystem

This Markov model can be represented by a 4×4 square matrix. The matrix tells us the probability of moving from one state to another (Figure 5-20).

Note that the matrix contains all the information of the model. One can tell the probability of moving from any state to any other state. To solve the Markov model for time dependent state probabilities, we assume that the system starts in the operating state, state 0 at time = 0. This is equivalent to assuming that the system is working properly when first installed and commissioned — a very realistic assumption.

If we assume that all the equipment is working at the beginning of our mission, we calculate state probabilities by matrix multiplication as shown in Figure 5-22.

$$\begin{array}{c c} & \text{To:} \quad 0 \quad\quad\quad 1 \quad\quad\quad 2 \quad\quad 3 \\ \text{From:} & \\ 0 & \begin{bmatrix} 1-(\lambda_{MV}+4\lambda_{SV}) & \lambda_{MV} & 4\lambda_{SV} & 0 \\ 0 & 1-(4\lambda_{SV}) & 0 & 4\lambda_{SV} \\ 0 & 0 & 1-(\lambda_{MV}) & \lambda_{MV} \\ 0 & 0 & 0 & 1 \end{bmatrix} \\ 1 & \\ 2 & \\ 3 & \end{array}$$

Figure 5-20. Valve Subsystem Transition Matrix

EXAMPLE 5-10

Problem: Solve the Markov model for the unreliability of the valve subsystem for a one-year mission time. The main valve has one failure mode and a failure rate of 0.000003 failures per hour. The secondary valves have one failure mode and a failure rate of 0.000002 failures per hour.

Solution: The Markov model is shown in Figure 5-19. The transition matrix is shown in Figure 5-20. Substituting the given failure rates into the transition matrix produces a numeric transition matrix shown in Figure 5-21.

$$\begin{array}{c c} \text{Markov} & \text{To:} \quad 0 \quad\quad\quad 1 \quad\quad\quad 2 \quad\quad\quad 3 \\ \text{From:} & \\ 0 & \begin{bmatrix} 0.999989 & 0.000008 & 0.000003 & 0 \\ 0 & 0.999997 & 0 & 0.000003 \\ 0 & 0 & 0.999992 & 0.000008 \\ 0 & 0 & 0 & 1 \end{bmatrix} \\ 1 & \\ 2 & \\ 3 & \end{array}$$

Figure 5-21. Valve Subsystem Numeric Transition Matrix

With a spreadsheet, one can simply copy and paste the matrix multiplication for as many times as needed. This provides a very simple way to obtain state probabilities as a function of time increments. The starting row, the first ten rows and the last ten rows of the spreadsheet are shown in Figure 5-23.

Since state 3 is the failure state, the probability of failure (unreliability) at a mission time of one year (8760 hours) is 0.00176. This number is lower

78   System Reliability Engineering

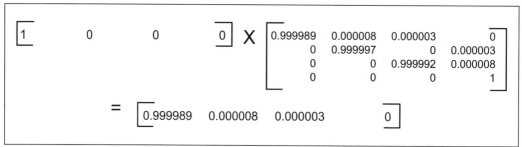

Figure 5-22. Matrix Multiplication at T = 1

0	1	0	0	0
1	0.999989	0.000008	0.000003	0
2	0.999978	1.6E-05	6E-06	4.8E-11
3	0.999967	2.4E-05	9E-06	1.44E-10
4	0.999956	3.2E-05	1.2E-05	2.88E-10
5	0.999945	4E-05	1.5E-05	4.8E-10
6	0.999934	4.8E-05	1.8E-05	7.2E-10
7	0.999923	5.6E-05	2.1E-05	1.008E-09
8	0.999912	6.4E-05	2.4E-05	1.344E-09
9	0.999901	7.2E-05	2.7E-05	1.728E-09
10	0.99989	7.999E-05	3E-05	2.16E-09

8751	0.9082265	0.0658621	0.0241596	0.0017518
8752	0.9082165	0.0658692	0.0241622	0.0017522
8753	0.9082065	0.0658763	0.0241647	0.0017526
8754	0.9081965	0.0658833	0.0241672	0.001753
8755	0.9081865	0.0658904	0.0241698	0.0017533
8756	0.9081765	0.0658975	0.0241723	0.0017537
8757	0.9081665	0.0659045	0.0241748	0.0017541
8758	0.9081565	0.0659116	0.0241774	0.0017545
8759	0.9081465	0.0659187	0.0241799	0.0017549
8760	0.9081365	0.0659257	0.0241824	0.0017553

Figure 5-23. Matrix Multiplication for the Mission Time

than the number obtained from the fault tree, 0.00184, as the Markov solution does not use the approximation techniques.

Realistic Safety Instrumented System Modeling

Most analysts doing safety instrumented system modeling use either fault trees or Markov models. Both methods provide a clear way to express the reality of multiple failure modes. Both methods, however, require careful modeling and appropriate solution techniques. Realistic levels of detail that include component failure modes, the effects of diagnostics, common cause failures, proof test effectiveness, and other variables are necessary. These topics are covered in the ensuing chapters.

Exercises

5-1. For a gas burner to work without failure, one of two identical fans, the burner, and the controls have to function without failure. The reliability of a fan is 0.9. The reliability of the burner is 0.98. The reliability of the control system is 0.85. The aforementioned reliabilities are for a period of five years. It can be assumed that the failure of each of the listed components is independent of any other failure.

Calculate the reliability of the furnace system over a five-year period.

5-2. A system has a series reliability block diagram as follows:

Failure rates are:

$\lambda_{AC\ POWER} = 0.002$ failures per year
$\lambda_{DC\ POWER\ SUPPLY} = 0.004$ failures per year
$\lambda_{LEVEL\ SWITCH} = 0.02$ failures per year
$\lambda_{TIME\ DELAY\ RELAY} = 0.025$ failures per year
$\lambda_{SOLENOID} = 0.08$ failures per year

What is the failure rate of the system in failures per year?

a. 0.131
b. 0.125
c. 0.129
d. 3.2 E-10
e. 0.869

5-3. What is the reliability of the system of Exercise 2 for a 1-year time interval?

a. 0.869
b. 0.131
c. 0.125
d. 0.9869
e. 0.877

5-4. A system has a the reliability block diagram shown below.

Availability of AC power = 0.99, Availability of the motor = 0.98

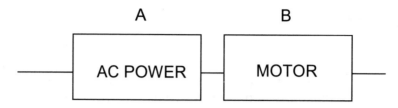

The system availability is closest to:
a. 1.97
b. 0.97
c. 0.98
d. 0.99
e. 0.9

5-5. A system has four components with the reliability block diagram shown below. If the power supply has a one-year reliability of 0.95 and the controller has a one-year reliability of 0.99, what is the system reliability for one year?

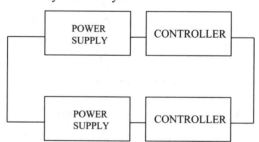

a. 1.881
b. 0.99646
c. 0.99965
d. 0.0035
e. 0.969

5-6.　Based on the system as shown in the figure above, draw a fault tree to represent a nuisance trip due to the failure of TX1/2 subsystem, i.e. the top event to be "Nuisance trip due to failure of transmitter subsystem." Transmitter voting is 1oo2. The following basic events are to be used:

Safe failure of TX1
Safe failure of TX2
Loss of power to transmitter

Calculate the probability of a nuisance trip using the following data:

Probability of safe failure for a one-year interval of TX1 = 0.02
Probability of safe failure for a one-year interval of TX2 = 0.02
Probability of loss of power for a one-year interval to transmitters = 0.01

5-7.　What is the P matrix for the Markov model of Figure 5-18?

5-8.　Draw a Markov model for the system of Figure 5-6.

REFERENCES AND BIBLIOGRAPHY

1. Goble, W. M., *Control Systems Safety Evaluation and Reliability*, second edition, NC: Research Triangle Park: ISA, 1998.

2. Billinton, Roy and Allan, R. N., *Reliability Evaluation of Engineering Systems: Concepts and Techniques*, NY: New York, Plenum Press, 1983.

3. Dhillon, B. S., *Reliability Engineering in Systems Design and Operation*, NY: New York, Van Nostrand Reinhold, 1983.

4. Lewis, E. E., *Introduction to Reliability Engineering*, NY: New York, John Wiley and Sons, Inc., 1994.

5. ISA-TR84.00.02-2002, Technical Report, *Safety Instrumented System (SIS) – Safety Integrity Level (SIL) Evaluation Techniques*, NC: Research Triangle Park, ISA, 2002.

6. IEC 61025, *Fault Tree Analysis*, Geneva: Switzerland, First Edition, 1990.

7. Henley, E. J., and Kumamoto, H., *Probabilistic Risk Assessment: Reliability Engineering, Design, and Analysis*, NJ: Piscataway, IEEE Press, 1992.

8. Brombacher A. C., "Fault Tree Analysis; why and when," *Advances in Instrumentation and Control - Anaheim*, ISA, 1994.

9. Rouvroye, J. L.; Goble, W. M.; Brombacher, A. C.; Spiker, R. Th. E., "A comparison study of qualitative and quantitative analysis techniques for the assessment of safety in industry," *Proceedings of PSAM III, International Conference on Probabilistic Safety Assessment and Management*, Crete, Greece, 1996.

6
Equipment Failure Modes

Introduction

A reliability engineer's first design priority is successful operation. Great effort must be made to ensure that things work. This priority is certainly logical for most systems as failure mode is not relevant.

In safety instrumented systems, however, the failure mode is very important. It makes a difference if the system fails and causes a false trip versus a failure that prevents the automatic protection.

Actual failures of instruments can be classified as "fail-safe," "fail-danger," or another failure mode. Such failure modes will be defined in this chapter in the context of an individual instrument. Note that sometimes the application must be understood before these classifications can be made. It must be remembered that the safety instrumented function may or may not fail when one instrument has failed. A redundant architecture may compensate for instrument failures.

Equipment Failure Modes

Instrumentation equipment can fail in different ways. We call these "failure modes." Consider a two-wire pressure transmitter. This instrument is designed to provide a 4 – 20 milliamp signal in proportion to the pressure input. Detailed failure modes, effects, and diagnostic analyses of several of these devices reveal a number of failure modes: frozen output, current to upper limit, current to lower limit, diagnostic failure, communications failure, and drifting/erratic output among perhaps others. These instrument failures can be classified into failure mode categories when the application is known.

If a single transmitter (no redundancy) were connected to a safety PLC programmed to trip when the current goes up (high trip), then the instrument failure modes could be classified as shown in Table 6-1.

Table 6-1. Transmitter Failure Mode Categories

Instrument Failure Mode	SIF Failure Mode
Frozen output	Fail-Danger
Output to upper limit	Fail-Safe
Output to lower limit	Fail-Danger
Diagnostic failure	Annunciation
Communication failure	No Effect
Drifting / erratic output	Fail-Danger

Consider possible failure modes of a PLC with a digital input and a digital output; both in a de-energize to trip (logic 0) design. The PLC failure modes can be categorized relative to the safety function as shown in Table 6-2.

Table 6-2. PLC Failure Mode Categories

Instrument Failure Mode	SIF Failure mode
Input stuck High	Fail-Danger
Input stuck low	Fail-Safe
Input circuit oscillates	Fail-Danger*
Output stuck high	Fail-Danger
Output stuck low	Fail-Safe
Improper CPU execution	50% Fail-Safe
	50% Fail-Danger
Memory transient failure	50% Fail-Safe
	50% Fail-Danger
Memory permanent failure	50% Fail-Safe
	50% Fail-Danger
Power supply low (out of tolerance)	Fail-Danger*
Power supply high (out of tolerance)	Fail-Danger*
Power supply zero	Fail-Safe
Diagnostic timer failure	Annunciation
Loss of communication link	No Effect
Display panel failed	No Effect
* unpredictable - assume worst case	

Final element components will fail also, and again the specific failure modes of the components can be classified into relevant failure modes depending on the application. It is important to know whether a valve will open or close on trip. Table 6-3 shows an example failure mode classification based on a close to trip configuration.

Table 6-3. Final Element Failure Mode Categories

Instrument Failure Mode	SIF Failure mode
Solenoid plunger stuck	Fail-Danger
Solenoid coil burnout	Fail-Safe
Actuator shaft failure	Fail-Danger*
Actuator seal failure	Fail-Safe
Actuator spring failure	Fail-Danger
Actuator structure failure - air	Fail-Safe
Actuator structure failure - binding	Fail-Danger*
Valve shaft failure	Fail-Danger*
Valve external seal failure	No Effect
Valve internal seal damage	Fail-Danger
Valve ball stuck in position	Fail-Danger
* unpredictable - assume worst case	

It should be noted that the above failure mode categories apply to an individual instrument and may not apply to the set of equipment that performs a safety instrumented function because the equipment set may contain redundancy. It should be also made clear that the above listings are not intended to be comprehensive or representative of all component types.

Fail-Safe

Most practitioners define "Fail-Safe" for an instrument as **a failure that causes a "false or spurious" trip of a safety instrumented function unless that trip is prevented by the architecture of the safety instrumented function.** Many formal definitions have been attempted that include "a failure which causes the system to go to a safe state or increases the probability of going to a safe state." This definition is useful at the system level and includes many cases where redundant architectures are used.

IEC 61508 uses the definition "failure which does not have the potential to put the safety-related system in a hazardous or fail-to-function state." This definition includes many failures that do not cause a false trip under any circumstances and is quite different from the definition practitioners need to calculate the false trip probability. Using this definition, all failure modes that are NOT dangerous are called "safe." This definition is not used in this book as most practitioners require more detail.

Fail-Danger

Many practitioners define "Fail-Danger" as **a failure that prevents a safety instrumented function from performing its automatic protection function.** Variations of this definition exist in standards. IEC 61508 provides a definition similar to the one used herein, which reads: "failure which has the potential to put the safety-related system in a hazardous or fail-to-function state." The definition from IEC 61508 goes on to add a note: "Whether or not the potential is realized may depend on the channel architecture of the system; in systems with multiple channels to improve safety, a dangerous hardware failure is less likely to lead to the overall dangerous or fail-to-function state." The note from IEC 61508 recognizes that a definition for a piece of equipment may not have the same meaning at the safety instrumented function level or the system level.

Annunciation

Some practitioners recognize that certain failures within equipment used in a safety instrumented function prevent the automatic diagnostics from correct operation. When reliability models are built, many account for the automatic diagnostics ability to reduce the probability of failure. When these diagnostics stop working, the probability of dangerous failure or false trip is increased. While these effects may not be significant, unless they are modeled, the effect is not known.

An annunciation failure is therefore defined as **a failure that prevents automatic diagnostics from detecting or annunciating that a failure has occurred inside the equipment.** Note that the failure may be within the equipment that fails or inside an external piece of equipment designed for the purpose of automatic diagnostics. These failures would be classified as "Fail-Safe" in the definition provided in IEC 61508.

No Effect

Some failures within a piece of equipment have no effect on the safety instrumented function, nor cause a false trip, nor prevent automatic diagnostics from working. Some functionality performed by the equipment is impaired, but that functionality is not needed. These may simply be called "No Effect" failures. They are typically not used in any reliability model intended to obtain probability of a false trip or probability of a fail-danger. Per IEC61508, these would be classified as "Fail-Safe" or may be excluded completely from any analysis depending on interpretation of the analyst.

Detected/Undetected

Failure modes can be further classified as "detected" or "undetected" by automatic diagnostics. In this book the classification is done at the instrument level, and the specific diagnostics are automatically performed somewhere in the safety instrumented system.

SIF Modeling of Failure Modes

When evaluating safety instrumented function safety integrity, an engineer must examine more than the probability of successful operation. The failure modes of the system must be individually calculated. The normal metrics of reliability, availability, and MTTF only suggest a measure of success. Additional metrics to measure safety integrity include probability of failure on demand (PFD), average probability of failure on demand (PFDavg), risk reduction factor (RRF), and mean time to fail dangerously (MTTFD). Other related terms are probability of failing safely (PFS) and mean time to fail spuriously (MTTFS).

PFS/PFD

There is a probability that a safety instrumented function will fail and cause a spurious/false trip of the process. This is called probability of failing safely (PFS). There is also a probability that a safety instrumented function will fail such that it cannot respond to a potentially dangerous condition. This is called probability of failure on demand (PFD).

PFDavg

PFD average (PFDavg) is a term used to describe the average probability of failure on demand. PFD will vary as a function of the operating time interval of the equipment. It will not reach a steady state value if any periodic inspection, test, and repair is done. Therefore, the average value of PFD over a period of time can be a useful metric if it assumed that the potentially dangerous condition (also called hazard) is independent from equipment failures in the safety instrumented function.

The assumption of independence between hazards and safety instrumented function failures seems very realistic. (NOTE: If control functions and safety functions are performed by the same equipment, the assumption may not be valid! Detailed analysis must be done to insure safety in such situations, and it is best to avoid such designs completely.) When hazards and equipment are independent, it is realized that a hazard may come at any time. Therefore, international standards have specified that PFDavg is an appropriate metric for measuring the effectiveness of a safety instrumented function.

PFDavg is defined as the arithmetic mean over a defined time interval. For situations where a safety instrumented function is periodically inspected and tested, the test interval is correct time period. Therefore:

$$PFDavg(TI) = \frac{1}{TI} \int_0^{TI} (PFD) dt \qquad (6\text{-}1)$$

This definition is used to obtain numerical results in several of the system modeling techniques. In a discrete time Markov model using numerical solution techniques, a direct average of the time dependent numerical values will provide the most accurate answer. When analytical equations for PFD are obtained using a fault tree, the above equation can be used to obtain equations for PFDavg (See Appendix F, System Architectures, for examples).

Exercises

6-1. A solenoid is normally energized in normal process operation. It is de-energized when a dangerous condition is detected and vents air from a pneumatic actuator. If the solenoid coil fails short circuit and burns out, the solenoid will de-energize. How should this failure mode be classified?

6-2. A set of equipment used in a safety instrumented function is non-redundant (1oo1). The total dangerous detected failure rate is 0.002 failures per year. The total dangerous undetected failure rate is 0.0005 failures per year. Restore time average is 168 hours. The equipment is inspected and tested every two years with 100% test coverage. What is the PFD? What is the PFDavg?

6-3. The failure rate (λ) for a pressure transmitter is 1.2×10^{-6} f/hr. The safe failure mode split is 50%. What is the dangerous failure rate?

6-4. A valve is designed to close on trip in a safety instrumented function. If this valve had a failure where internal seals were damaged and could not completely stop flow, how would this failure be classified?

6-5. A flame detector used in a burner management application falsely indicates a flame when there is none. How would that failure mode be classified?

6-6. A gas detector used in a flammable gas shutdown function falsely indicates the presence of flammable gas. How would that failure mode be classified?

6-7. A safety PLC communicates shutdown status to the operator via a communication link. If this link fails to communicate, how would that failure mode be classified?

6-8. A safety instrumented function has a valve energized and open. The valve must close when a demand is detected. The valve is fitted to a piston type pneumatic actuator that has an O-ring seal around the piston. This seal degrades with time and gets sticky. If it is left in position a long period of time, it will cause the actuator to stick in place. How would this failure mode be classified?

7

SIF Verification Process

The Conceptual Design Process

The conceptual design process for a safety instrumented function begins with a Safety Requirements Specification (SRS). The SRS is an important document. It should contain complete specifications for designing all the safety instrumented functions. For each SIF the following information should be included:

- the hazard and its consequences
- the demand frequency for the hazard
- a reference to the appropriate P&ID drawing
- a definition of the process safe state
- a description of the safety instrumented function
- a description of the process measurements and trip points
- a description of the output response required for both primary equipment and auxiliary/secondary equipment
- the relationship between process measurements and outputs, including logic, mathematical functions and any required permissive – this must be specified for all modes of operation, e.g., startup, normal, abnormal, emergency, shutdown, etc.
- the required safety integrity level
- target proof test intervals
- maximum allowable spurious trip rate
- maximum response time requirement for the SIF
- requirements for manual activation of the SIF
- requirements for reset of the SIF (latching or automatic reset)

- SIF response to diagnostic faults (automatic shutdown, alarm only or other)
- requirements for human interface – What variables must be displayed? What variables must be input?
- maintenance override capability requirements
- estimates for mean time to restore, startup time after a trip, etc.
- expected environmental conditions during normal operation and emergency situations

Other items may be required for specific applications including any local regulatory requirements (must meet NFPA85 [Ref. 1] for example), references to company specific requirements or other requirements.

Given the design objectives in the SRS, the designer must choose equipment, determine if redundancy is needed, determine SIF testing techniques and perform a set of calculations to determine if various metrics are within the range for the desired SIL level. A diagram of the process is shown in Figure 7-1.

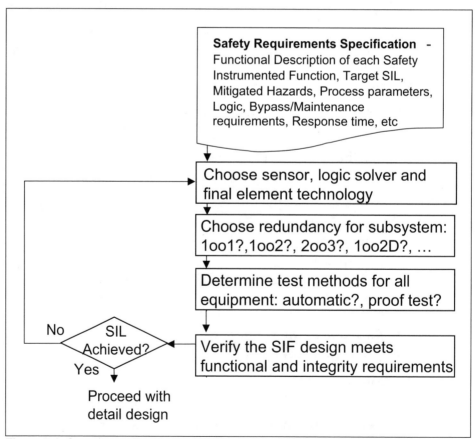

Figure 7-1. Conceptual Design Process

It is during this conceptual design step that probabilistic calculations are done to show that any given design meets the performance requirements of the SIL level. These calculations are the key to a performance based standard like IEC 61508 (Ref. 2) or ANSI/ISA 84.00.01-2004 (IEC 61511) (Ref. 3). This performance-based approach was done as an alternative to a set of very prescriptive rules that embody specific design practices. Instead of listing what each designer must do in terms of specific equipment choices and levels of redundancy, the engineer doing the conceptual design is given the freedom to choose equipment, choose redundancy levels and make tradeoffs between equipment types, testing and even cost. The conceptual design process must be taken seriously. Because with the freedom of a performance based standard, comes the responsibility to properly model the probabilistic performance of the design.

The standards allow any internationally accepted methodology for the purpose of performing the probabilistic calculations and the most common methods are described in this book -- simplified equation, fault trees and Markov models.

Equipment Selection

Equipment used in a safety instrumented function must be carefully chosen. The instrumentation must be fully capable of performing the functional requirement. All equipment must be justified so that the end user is totally confident that the instrumentation will properly perform in the intended application.

Materials used in the instruments must be compatible with process materials if the instrumentation sees process wetted service. Process environmental conditions must not exceed the instrumentation ratings. The functional safety of the instrument must be assessed. All justification decisions must be documented as part of project records.

Special attention must be paid to the support systems; e.g. power supplies, air supplies, wiring methods. For energize to trip designs, these systems become safety critical since they have a direct impact on safety.

Equipment Functional Safety Assessment

ANSI/ISA-84.00.01-2004 (IEC 61511 Mod) requires that equipment used in safety instrumented systems be chosen based on either IEC 61508 certification to the appropriate SIL level or justification based on "prior use" criteria (ANSI/ISA-84.00.01-2004 (IEC 61511Mod), Part 1, Section 11.5.3). However the ANSI/ISA-84.00.01-2004 (IEC 61511 Mod) standard does not give specific details as to what the criteria for "prior use" means. Most agree however that if a user company has many years of documented successful experience (no dangerous failures) with a particular version of a particular instrument this can provide justification for using that instrument even if it is not safety certified.

Most agree also that prior use requires that a system be in place to record all field failures and failure modes at each end user site. Version records of the instrument hardware and software must be kept as significant design changes may void prior use experience. Operating conditions must be recorded and must be similar to the proposed safety application.

Many users have asked instrument manufacturers to help with justifications. In turn, manufacturers have had different levels of assessment performed by third party assessors. Such assessments are normally done by third-party experts like exida, TÜV or FM. Often, two or more assessor companies will team together to do the assessment as requested by the instrument manufacturer.

Assessments can help reduce the burden of documentation when an end user attempts to justify an instrument for use on safety applications. In the marketplace three basic levels of assessment have been done on instrumentation products with variation between assessment companies.

FMEDA Assessment of an Instrument

A hardware analysis called a Failure Modes Effects and Diagnostics Analysis (FMEDA) can be done to determine the failure rates and failure modes of an instrument (Ref. 4). This is done to provide the safety design engineer with the data needed to more accurately perform probabilistic analysis.

A FMEDA is a systematic detailed procedure that is an extension of the classic FMEA procedure developed and proven decades ago. The technique was first developed for electronic devices and recently extended to mechanical and electro-mechanical devices (Ref. 5). This analysis for hardware devices provides the required failure data needed for SIF verification.

Some assessors also do a useful life analysis to provide the safety instrumentation engineer with knowledge of any wear out mechanisms and the time periods until wear out. Preventative maintenance programs can be established to replace instruments at the end of their useful life knowing this information. Some FMEDA analyses are also extended to evaluate the effectiveness of proposed proof test methods. This provides the safety instrument engineer with proof test coverage factors used for more realistic PFH/PFD/PFDavg calculations.

Note that an FMEDA alone should not be considered sufficient justification for selecting a product.

Prior Use Assessment of an Instrument

The primary function of Prior Use assessment is to obtain a level of comfort regarding systematic (design) faults. One must develop confidence that a product does not have inherent design flaws. Sometimes the data gathered in a Prior Use assessment is used to calculate failure rates but this must be approached with caution whenever it is suspected that all failures are not reported.

Prior use assessment is the responsibility of an end user. However some manufacturers are working to help their customers in prior use evaluations by providing information. Some manufacturers have had a third party assessment of the field failure records that exist for a device. Failures attributable to both hardware and software are considered. As part of this process an assessment must also be made of the field return data gathering procedures and the product modification process.

Sufficiently detailed procedures must exist within a manufacturer to insure a reasonable level of quality in the data. The field data must show environmental limitations and application limitations. Since the manufacturer does not actually "use" the equipment, this information must be obtained from end users' data. Methods for collecting, reporting and analyzing this data should be critically reviewed. Typically field failure records are used to calculate failure rates and these are compared to the failure rates from a FMEDA. If they are significantly lower than the FMEDA numbers there is evidence that significant design flaws do not exist. Field failure rate numbers from this data should not be used for SIL verification. Remember that field failures are notoriously under-reported to the manufacturer.

All field failure data must be reviewed keeping in mind design revisions. When significant design changes are made, then field experience with previous designs should not be considered.

An end user may consider the data provided by the manufacturer but it is much better if the end user has their own detailed failure records that provide statistically significant failure information about a product.

Full Assessment according to IEC 61508

A complete IEC 61508 assessment includes a FMEDA, a study of Prior Use and adds an assessment of all fault avoidance and fault control measures during hardware and software development as well as detail study of the testing, modification, user documentation and manufacturing processes. The objective of all this effort is to provide a high level of assurance that an instrument has sufficient quality and integrity for a safety instrumented system application. This is clearly more important for products containing software as many end users have the strong opinion that software is "bad for safety systems." This attitude comes from experience where products have failed due to software faults.

A full assessment according to IEC 61508 is the most comprehensive functional safety assessment available. It is becoming more necessary as software content grows in our instrumentation equipment. Field failures due to design faults (systematic errors) are increasing. These are mostly software faults. This type of failure is very unlikely to be reported to the manufacturer, as "repair" is often a "software reset" or power cycle. This type of failure is often not even recorded in the end user maintenance system since no "replacement" needs to be made. Hence "prior use" or field failure evaluation techniques based on returns to the manufacturer are not sufficiently effective and very few end users have the failure recording systems needed to assure a high level of integrity.

Many of the requirements of IEC 61508 focus on the elimination of systematic faults. In order to demonstrate compliance with all requirements of IEC 61508, the design and development process used to create an instrument must show extensive use of many techniques for "fault control" and "fault avoidance." The IEC 61508 standard defines a set of practices that represent good software and hardware engineering. Most experts believe that these methods are the best techniques available to provide high design quality.

Another benefit of having instrumentation products that are 61508 certified is that they have demonstrated higher levels of design quality and software quality. Given the number of failures due to software in instrumentation equipment, this is a top priority for many end users. It should be noted that the quality within a product is directly proportional the to "SIL Capability" rating assigned to the product. A SIL3 capable product is generally considered to have an order of magnitude higher quality than a SIL2 capable product.

Full compliance with the requirements of IEC 61508 is seen when a product does not have any significant "restrictions" on usage as documented in the product "Safety Manual." A large safety manual with a long detailed list of instructions on how to make the product "safe" is a sure sign the manufacturer does not meet requirements unless these restrictions are implemented by the end user.

The primary differences in the assessment techniques are summarized in Table 7-1.

Redundancy

After equipment is selected, the next step in the conceptual design process is the decision to use multiple instruments to serve the same purpose – redundancy. Redundancy is configured to provide continued system operation even though one or more specific instruments may fail – fault tolerance. Some redundant architectures provide fault tolerance against a false trip. Some redundant architectures provide fault tolerance against a dangerous failure and some architectures can provide fault tolerance against multiple failure modes. Detailed explanations of various redundant architectures can be found in Appendix F.

Table 7-1. Assessment Techniques

Assessment Criteria	FMEDA only	Prior Use	IEC 61508 Certification
Detail analysis of hardware failure modes	X		X*
Detail analysis of hardware diagnostic capability	X		X*
Analysis of hardware useful life	X*		X*
Analysis of proof test effectiveness	X*		X*
Assessment of operational hours based on manufactured units		X	X
Assessment of Configuration Management system per requirements of IEC 61508		X	X
Assessment of Field Failure Return System - field failures corrected		X*	X
Assessment of Field Failure Return System - notification to users of safety issues		X*	X
Assessment of design revision history - few revisions based on design faults		X*	X
Assessment of hardware design process			X
Assessment of hardware testing techniques			X
Assessment of software requirements			X
Assessment of software criticality			X
Assessment of software design techniques			X
Verification of Safety Manual per IEC 61508			X
Assessment of software testing techniques			X
Assessment of product testing techniques including environmental testing			X
Assessment of manufacturing process			X

* Depends on assessment agency - not all third-party agencies perform the same analysis

SIF Testing Techniques

The safety and availability of a set of equipment used for a safety instrumented function may benefit from testing. However, that depends on redundancy and how often the demand occurs. Three modes of operation have been defined in IEC 61508 for equipment providing a safety instrumented function: continuous demand mode, high demand mode and low demand mode. This book will use the IEC 61508 definitions to designate those three different situations.

These three modes have been defined because SIF testing may or may not be given credit depending on the level of redundancy and the mode. The probability of failure on demand is calculated differently for each mode. The essential differences are due to the relationship between the dangerous condition (the demand) and the diagnostic testing.

Three time intervals must be known to define what credit may be taken for automatic diagnostic testing and manual proof testing. These three time intervals are the average demand interval, the manual proof test interval and the automatic diagnostic test interval (usually the worst-case time is considered). The three modes and their relationships are shown in Table 7.2.

Table 7-2. SIF Modes and Time Interval Relationships

Mode	Demand Interval versus Automatic Diagnostic Interval	Demand Interval versus Manual Proof Test Interval	Probability Measure
Continuous Demand	DI <= ATI	DI <= PTI	PFH
High Demand	DI >> ATI	DI <= PTI	PFH
Low Demand	DI >> ATI	DI >> PTI	PFDavg

Continuous Mode

In continuous mode, the demand is effectively always present. Dangerous conditions always exist and a dangerous failure of the safety instrumented function will immediately result in an incident. There are no safety benefits that can be claimed for manual proof testing or even automatic on-line diagnostics in a single channel system (1oo1). By the time the diagnostics detect the fault and initiate action, it is too late. Therefore, in continuous demand mode probability evaluation cannot take credit for any diagnostics except in redundant systems.

The probability evaluation is done by comparing the calculated probability of failure per hour (PFH) against the PFH table shown in Figure 7-4.

High Demand Mode

In high demand mode, the dangerous condition is not always present but does occur frequently. We can define the "high demand" mode by stating that it occurs when a demand occurs nearly as often as or more often than any practical manual proof test interval but considerably slower than the automatic diagnostic test and response time. The exact demand rate is not important. It is the ratio of demand rate to manual proof test rate and automatic diagnostic and response rate that defines the region. The threshold between continuous demand and high demand depends on the ratio of the demand rate to the automatic test rate. The reason to distinguish these modes is that one may take credit for automatic diagnostics in high demand mode even in a single channel (1oo1) system.

If automatic diagnostics complete execution at a frequency two or more times the expected average demand rate and the system responds to a diagnostic fault by initiating a move to the safe state, then safety could be improved and some credit can be taken in the probability of failure modeling. If automatic diagnostics are not done many times faster than the demand rate then detailed probability models showing the demand probability and deterministic diagnostic time periods must be done. Repair time must be carefully modeled if the safety instrumented function is not programmed to automatically initiate safety action upon detecting a dangerous fault within the instrumentation equipment. This level of modeling is complicated and many choose to simply classify these safety instrumented functions in the continuous mode category.

If the automatic diagnostics complete execution many times faster than the average expected demand rate and automatic safety action is initiated, then the probability models can be simplified and the effects of diagnostics can be given full credit even in a single channel (1oo1) system. IEC 61508 suggests that a number of ten times be used (Part 2, 7.4.3.2.2, e, Note 3). In order to determine this, the time period of the automatic diagnostics must be known.

In high demand mode, the probability evaluation is done by comparing the calculated probability of failure per hour (PFH) against the PFH table shown in Figure 7-4.

Low Demand Mode

In low demand mode, a dangerous condition is expected very infrequently. The threshold for classification between high demand and low demand should be determined by the ratio of any planned manual proof test interval to the average interval between demands. IEC 61508 states that the proof test interval must be no greater than half the expected average demand interval or greater than one year (Part 4, Clause 3.5.12). Detailed probabilistic modeling will show that manual proof test diagnostics can and should be given credit when the average demand

> **EXAMPLE 7-1**
>
> **Problem:** A set of non-redundant (1oo1) equipment is used to implement a safety instrumented function. Within the equipment, automatic diagnostics complete execution every one second. The instrument is programmed to take the process to a safe state when an internal failure of the equipment is detected. A dangerous condition occurs every one minute on average. What is the mode of operation and can the automatic diagnostics be given credit in the probability of failure calculation?
>
> **Solution:** The automatic diagnostics perform their function sixty times during the average demand period and perform an automatic process shutdown. A detailed probability model showing the exact effect of deterministic automatic diagnostics is not necessary as it would show diagnostics are effective in improving safety. This SIF would be classified as high demand mode.

interval exceeds twice the manual proof test interval. Simplified equations that do not specifically account for the demand period will give an optimistic result unless the average demand period is ten times greater than manual proof test interval.

> **EXAMPLE 7-2**
>
> **Problem:** Layer of protection analysis has indicated that a demand would occur every 5 years on average for a particular process hazard. Although most automatic diagnostics execute every minute, the worst-case time period for automatic diagnostics within the equipment is once per week. A proof test interval of one year is proposed for a manual test and inspection. Would this SIF be classified as low demand?
>
> **SOLUTION:** Automatic diagnostics are performed many times within the expected average demand interval. The proof test is done at least two times within the expected average demand period so the SIF would be classified as low demand.

Fortunately in most situations in the process industries, especially when independent layers of protection (Ref. 6 and 7) are properly designed and considered in the demand rate analysis, average demand intervals are very high. Often the average demand interval will exceed one hundred years.

In low demand mode safety instrumented functions, the person performing the SIF verification calculations must:

1. define the proposed proof test procedures and

2. estimate the effectiveness of all procedures via an estimate of proof test diagnostic coverage.

That information is needed along with other parameters to perform an accurate average probability of failure on demand calculation.

> **EXAMPLE 7-3**
>
> **Problem:** A pressure transmitter is needed for a safety instrumented function. What proof test procedures should be performed and what coverage factors should be used for those procedures?
>
> **Solution:** Many manufacturers recommend proof test procedures for low demand safety instrumented system applications. The information is found in the "safety manual." That document may be part of another manual or may be a separate document. Referring to the safety manual section of a pressure transmitter (Ref. 4), proof test options with associated coverage factors are given (Figure 7-2). The test titled "Five Year Proof Test" has a manual proof test coverage of 65%. The test titled "Ten Year Proof Test" has a manual proof test coverage of 99%.

Reliability and Safety Metric Calculation

Probabilistic calculations are done to determine if the design meets the safety integrity requirement after:

1. equipment is chosen,

2. redundancy designs are completed and

3. the automatic test capabilities / manual proof testing goals are established based on SIF demand mode,

The calculations may be done with simplified equations, fault trees, Markov models or other techniques depending on the complexity of the model and the demand mode of operation.

SIF Identification

The first step in the calculation process is to properly identify the equipment required for each safety instrumented function. All equipment associated with a particular SIF must be classified into "primary" – equipment needed to provide the required protection against the identified hazard and "auxiliary" – equipment that provides useful functionality but not required to protest against the hazard. This classification is important because only primary equipment is included in the PFDavg analysis and the SFF analysis.

Quick Installation Guide
Preliminary-
Rosemount Proprietary Information   **Rosemount 3051S Series**

SAFETY INSTRUMENTED SYSTEMS

Additional Safety Instrumented Systems information is available in the Rosemount 3051S Series Pressure Transmitter reference manual (document number 00809-0100-4801). The manual is also available electronically on www.rosemount.com or by contacting an Emerson Process Management representative.

Installation
The 3051S SIS can be used in safety instrumented functions up to SIL2 with a single instrument and SIL3 with hardware fault tolerance of 1. No special installation is required. Follow standard installation practices as outlined in this document.

Set-up
The security switch must be in the ON position during normal operation.

Configuration
No special configuration is required. Verify configuration as outlined in this document. Use any HART-compliant master to communicate with and configure the 3051S SIS. DD revision 3051S SIS Dev. 1 Rev 1 is required.

Operation and Maintenance
Proof Test and Inspection:

The following proof tests are recommended. Proof test results must be documented with corrective actions taken in the remote possibility that safety functionality is not correct.

To perform a HART Loop Test follow the "Loop Test" instructions available in the 3015S reference manual (document number 00809-0100-4801).

Five-Year[1] Proof-Test
This proof test, when combined with a HART Loop Test, will detect more than 65% of DU failures not detected by the automatic diagnostics programmed into the 3051S SIS.
1. Enter the milliampere value representing a high alarm state
2. Check the reference meter to verify the mA output corresponds to the entered value.
3. Enter the milliampere value representing a low alarm state
4. Check the reference meter to verify that the mA output corresponds to the entered value.
5. Execute the Master Reset command to initiate start-up diagnostics.

Ten-Year[1] Proof-Test
This proof test, when combined with a HART Loop Test, will detect over 98% of DU failures not detected by the automatic diagnostics programmed into the 3051S SIS.
1. Perform a minimum two point calibration check using the 4-20mA range points as the calibration points.
2. Check the reference mA meter to verify the mA output corresponds to the pressure input value.
3. If necessary, use one of the "Trim" procedures available in the 3051S reference manual to calibrate.
4. Execute the Master Reset command to initiate start-up diagnostics.

(1) May be a longer proof test interval as justified by PFDavg calculation.

Figure 7-2. Safety Manual Page (reprinted with permission of Rosemount)

SIF Verification Process 101

EXAMPLE 7-4

Problem: A safety instrumented function is identified in a SRS. If a low liquid level is detected in a separation unit, the outlet valve must be closed to protect downstream equipment from high pressure "blow-by" which is the identified hazard. The inlet valve must also be closed, a pump must be turned off to avoid pump damage and the inlet valve for another process unit must be turned off to minimize process disruption. The logic for this function is given in a cause and effect diagram shown in Figure 7-3. What equipment is classified as primary versus auxiliary?

Solution: For each piece of equipment related to the safety instrumented function, one must ask if that equipment is needed to protect against the specified hazard. In this SIF, the hand-switch was added only to meet local regulatory requirements and is not part of the automatic protection so it is excluded. The pump is turned off to protect it from overload so it is not part of this SIF. The inlet valve for the other unit does not have to close to protect against this hazard so it is excluded. Although the need for the inlet valve closure is debatable, it does help reduce downstream pressure and was therefore included in the SIF. The SIF primary equipment is the LT-2025 level sensor, the VI-2002 Inlet Valve and the VI-2003 Outlet Valve. This is marked in the cause and effect diagram with an X. Other equipment is auxiliary. It is marked in the cause and effect diagram with an A. This information must be documented in the Safety Requirements Specification (SRS).

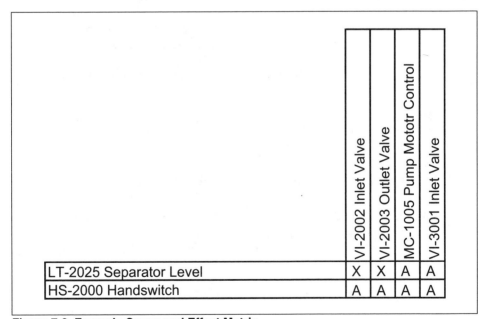

Figure 7-3. Example Cause and Effect Matrix

Continuous / High Demand Mode Verification Calculation

Remember that in continuous demand mode credit can be given for automatic diagnostics or for proof test procedures only for redundant systems. In addition, many of the calculation assumptions made for low demand mode do not apply. The "probability of dangerous failure per hour" must be calculated based on all dangerous failures.

This number is compared to the Continuous/High Demand probability chart from IEC 61508 to determine the SIL achieved by the design. This chart is shown in Figure 7-4.

Safety Integrity Level	Probability of dangerous failure per hour (Continuous mode of operation)
SIL 4	$>=10^{-9}$ to $<10^{-8}$
SIL 3	$>=10^{-8}$ to $<10^{-7}$
SIL 2	$>=10^{-7}$ to $<10^{-6}$
SIL 1	$>=10^{-6}$ to $<10^{-5}$

Figure 7-4. Continuous Demand Mode Dangerous Probability Limits per SIL

In high demand mode, credit may be given for automatic diagnostics even in a single channel (1oo1) system depending on the speed of the diagnostics and diagnostic response versus average demand interval.

Low Demand Mode Verification Calculation

When a proof test is done at least twice during an expected average demand period, low demand mode is appropriate and the calculated result can be done with PFDavg as defined in Chapter 4. The important variables that must be considered in the calculation include:

1. failure rates per failure mode of all instruments,

2. diagnostic coverage of automatic diagnostics,

3. mission time which should be the time interval between process unit overhaul or demand interval, whichever is shorter,

4. time interval between proof tests,

SIF Verification Process   103

> **EXAMPLE 7-5**
>
> **Problem:** A set of non-redundant (hardware fault tolerance = 0) safety equipment is used to perform a safety instrumented function in continuous demand mode. Diagnostic time is given as one second. The following failure rate data is obtained when adding the failure rates of the categories of all components:
>
> Lambda SD = 10×10^{-6} failures per hour, Safe Detected
>
> Lambda SU = 5×10^{-6} failures per hour, Safe Undetected
>
> Lambda DD = 8.5×10^{-6} failures per hour, Dangerous Detected
>
> Lambda DU = 0.5×10^{-6} failures per hour, Dangerous Undetected
>
> (NOTE: the terms detected and undetected refer to failures diagnosed by automatic diagnostics not those detected by the overt false trip of the SIF.)
>
> What SIL level is achieved by this design based on probability of dangerous failure per hour requirements?
>
> **Solution:** All dangerous failures will cause an incident because the dangerous condition is always present in the continuous mode. One second is not enough time to bring the process to a safe state. The total dangerous failure rate is 9×10^{-6} failures per hour. That meets the requirements for SIL1 per Figure 7-4.

5. proof test coverage factor,

6. average proof test time if the SIF is bypassed during the test,

7. common cause (beta factor) parameter estimates, and

8. average restore time(s).

The PFDavg may be calculated using one of several different techniques as described in Chapter 5. The most common approaches involve simplified equations, fault trees or Markov models. The resultant PFDavg value must be compared with a table from ANSI/ISA-84.00.01-2004 (IEC 61511 Mod) similar to Figure 7-5.

Architectural Constraints for Low Demand Mode

ANSI/ISA-84.00.01-2004 (IEC 61511 Mod) has a requirement for minimum levels of "hardware fault tolerance" as a function of SIL level. This means that redundancy for purposes of achieving the safety function must be done depending on the SIL level target of the SIF. For field instruments and non-programmable logic solvers, the chart is shown in Figure 7-6.

EXAMPLE 7-6

PROBLEM: A set of non-redundant (hardware fault tolerance = 0) safety equipment is used to perform a safety instrumented function in high demand mode. Diagnostic time is given as one second. The system is programmed to take the process to a safe state when a diagnostic indicates an internal failure. The response time of the system to achieve a safe state is 50 milliseconds. The process safety time is two seconds. An average demand interval is one minute. The following failure rate data is obtained when adding the failure rates of the categories of all components:

Lambda SD = 10×10^{-6} failures per hour, Safe Detected

Lambda SU = 5×10^{-6} failures per hour, Safe Undetected

Lambda DD = 8.5×10^{-6} failures per hour, Dangerous Detected

Lambda DU = 0.5×10^{-6} failures per hour, Dangerous Undetected

(NOTE: the terms detected and undetected refer to failures diagnosed by automatic diagnostics not those detected by the overt false trip of the SIF.)

What SIL level is achieved by this design based on probability of dangerous failure per hour requirements?

SOLUTION: The diagnostics operate rapidly and complete execution sixty times per expected demand period. The diagnostic test time plus the response time is within the process safety time. Therefore dangerous detected failures will be converted into safe failures. The remaining dangerous failure rate is 0.5×10^{-6} failures per hour. That meets the requirements for SIL2 per Figure 7-4.

Safety Integrity Level	Average probability of failure on demand (Low demand mode of operation)
SIL 4	>=10^{-5} to <10^{-4}
SIL 3	>=10^{-4} to <10^{-3}
SIL 2	>=10^{-3} to <10^{-2}
SIL 1	>=10^{-2} to <10^{-1}

Figure 7-5. Low Demand Mode Probability Chart

EXAMPLE 7-7

Problem: A set of non-redundant (1oo1) safety equipment is used to perform a safety instrumented function in low demand mode. The equipment is to be inspected and fully restored every five years. Therefore the manual proof test interval is five years and the manual proof test effectiveness can be assumed to be 100%.

When failures are detected by the automatic diagnostics, average restore time is 24 hours. The following failure rate data is obtained when adding the failure rates of the categories of all components:

Lambda SD = 5×10^{-6} failures per hour, Safe Detected

Lambda SU = 5×10^{-6} failures per hour, Safe Undetected

Lambda DD = 6.5×10^{-6} failures per hour, Dangerous Detected

Lambda DU = 0.5×10^{-6} failures per hour, Dangerous Undetected

(NOTE: the terms detected and undetected refer to failures diagnosed by automatic diagnostics not those detected by the overt false trip of the SIF.)

What SIL level is achieved by this design based on average probability of failure on demand?

Solution: The simplest approach is the use the simplified equations given in Appendix F. Although these provide only a rough approximation, they are quite useful for simple designs like this one. Since we will assume that proof test effectiveness is 100%, we can use equation F-2:

$$PFDavg_{1oo1} = \lambda^{DD} \times RT + \lambda^{DU} \times TI/2$$

$$= 6.5 \times 10^{-6} \times 24 + 0.5 \times 10^{-6} \times 8760 \times 5/2$$

$$= 0.000156 + 0.01095 = 0.011106$$

Based on the PFDavg chart of Figure 7-5, the design qualifies for SIL 1.

SIL	Minimum hardware fault tolerance
1	0
2	1
3	2
4	Special requirements apply (see IEC 61508)

Figure 7-6. IEC 61511 Minimum Hardware Fault Tolerance for Field Devices

When following this chart without an exception, the designer must use two transmitters for SIL2 and trip if either transmitter indicates a trip. When designing a SIL3 SIF without an exception, the designer must use three transmitters or three valves, each capable of performing the safety function.

The standard does allow the designer to take credit for one SIL level if a product is chosen on the basis of "prior use" and meets the following restrictions:

- the device allows adjustment of process-related parameters only, e.g., measuring range, upscale or downscale failure direction, etc.;
- the adjustment of the process-related parameters of the device is protected, e.g., jumper, password;
- the function has a SIL requirement less than 4.

If these restrictions are met, only one transmitter or valve is needed for a SIL2 SIF to meet this requirement. Alternatively the charts of IEC 61508 may be used for field devices. Given the lack of definition as to what "prior use" really means, the authors prefer to use the tables from IEC 61508 which are more flexible, provide at least the same level of "exception" for products with sufficient design quality and are clearly justifiable. The disadvantage of these charts is that the safe failure fraction must be calculated for the field devices.

For programmable electronic (PE) logic solvers another chart is presented in the ANSI/ISA-84.00.01-2004 (IEC 61511Mod) standard. It is shown in Figure 7-7.

SIL	Minimum Hardware Fault Tolerance		
	SFF < 60%	SFF 60% to 90%	SFF > 90%
1	1	0	0
2	2	1	0
3	3	2	1
4	Special requirements apply - See IEC 61508		

Figure 7-7. Constraint Table for PE Logic Solvers – IEC 61511

The use of this chart requires the calculation of a measure called the Safe Failure Fraction (SFF). This chart is completely equivalent to the same chart in IEC 61508.

Architecture Constraints from IEC 61508

IEC 61508 provides two charts that are used for all products, field instruments or logic solvers. For each product the Safe Failure Fraction (SFF) must be calculated. The SFF is defined as

$$\text{SFF} = \frac{\lambda^{SD} + \lambda^{SU} + \lambda^{DD}}{\lambda^{SD} + \lambda^{SU} + \lambda^{DD} + \lambda^{DU}} \tag{7-1}$$

It can be seen that this metric is a ratio of failure rates and not dependent on the total failure rate. The result is always a number between zero and one. A high number is good. It measures the natural tendency of an instrument to fail safety or detect dangerous failures.

If "%Safe" is defined to be

$$\%\text{Safe} = \frac{\lambda^S}{\lambda} \tag{7-2}$$

Then

$$\lambda^S = \%\text{Safe} \times \lambda \tag{7-3}$$

and

$$\lambda^D = (1 - \%\text{Safe}) \times \lambda \tag{7-4}$$

Given that

$$\lambda^S = \lambda^{SD} + \lambda^{SU}$$

and

$$\lambda^{DD} = C^D \times \lambda^D$$

$$\lambda^{DU} = (1 - C^D) \times \lambda^D$$

Substituting into Equation 7-1,

$$\text{SFF} = \%\text{Safe} + (1 - \%\text{Safe}) \times C^D \tag{7-5}$$

Equation 7-5 shows that the SFF is independent of total failure rate. This is an important characteristic as the metric was created to provide limits on SIF designs. The minimum levels of redundancy from the charts are required to protect against the use of optimistic failure rate data leading one to conclude that non-redundant designs would unrealistically achieve higher SIL levels. The SFF is generally applied to each subsystem in a SIF – sensor, logic solver and final element.

There are two categories of instrumentation equipment defined in IEC 61508 – Type A and Type B. A subsystem is classified as Type A "if, for the components required to achieve the safety function:

- the failure modes of all constituent components are well defined;
- the behavior of the subsystem under fault conditions can be completely determined; and
- there is sufficient dependable failure data from field experience to show that the claimed rates of failure for detected and undetected dangerous failures are met."

And practically speaking, Type B is everything else.

Examples of products typically classified as Type A include relays, solenoids, pneumatic boosters, actuators, valves and even simple electronic modules with resistors, capacitors, op amps, etc. Any smart product with a microprocessor or complex ASIC (Application Specific Integrated Circuit) is considered Type B. These are classified Type B because of their complex designs in combination with a relatively short operational history for any given generation. There is simply not enough time to gather sufficient operating history to develop a comprehensive understanding of possible systematic design errors, the resulting failure rates and failure modes. By the time enough experience starts to accumulate, a new generation of technology is introduced!

For Type A components, the minimum hardware fault tolerance chart per IEC 61508 is shown in Figure 7-8.

Safe Failure Fraction	Hardware Fault Tolerance		
TYPE A	0	1	2
< 60 %	SIL1	SIL2	SIL3
60 % < 90 %	SIL2	SIL3	SIL4
90 % < 99 %	SIL3	SIL4	SIL4
> 99 %	SIL3	SIL4	SIL4

Figure 7-8. Type A Architecture Requirements IEC 61508

> **EXAMPLE 7-8**
>
> **Problem:** A simple SIF was designed with a pressure switch hardwired to a two-way solenoid. The pressure switch opens on a high pressure demand and de-energizes the solenoid which will take the process to a safe state. This SIF has no automatic diagnostics; no complex microprocessors and both components are considered Type A. The failure rates are given below.
>
> Pressure Switch:
> > Lambda SD = 0×10^{-6} failures per hour, Safe Detected
> > Lambda SU = 2.4×10^{-6} failures per hour, Safe Undetected
> > Lambda DD = 0×10^{-6} failures per hour, Dangerous Detected
> > Lambda DU = 3.6×10^{-6} failures per hour, Dangerous Undetected
>
> (NOTE: the terms detected and undetected refer to failures diagnosed by automatic diagnostics not those detected by the overt false trip of the SIF.)
>
> Solenoid:
> > Lambda SD = 0×10^{-6} failures per hour, Safe Detected
> > Lambda SU = 3.8×10^{-6} failures per hour, Safe Undetected
> > Lambda DD = 0×10^{-6} failures per hour, Dangerous Detected
> > Lambda DU = 1.2×10^{-6} failures per hour, Dangerous Undetected
>
> (NOTE: the terms detected and undetected refer to failures diagnosed by automatic diagnostics not those detected by the overt false trip of the SIF.)
>
> According to the architecture limits of IEC 61508, to what SIL does this design qualify?
>
> **Solution:** The sensor subsystem consists of one switch, Type A. It has hardware fault tolerance of 0 since one dangerous failure will fail the SIF. The SFF is 40%. According to Figure 7-8. Type A Architecture Requirements IEC 61508, the subsystem qualifies for SIL 1.
>
> The final element subsystem consists of one solenoid, Type A. It has a hardware fault tolerance of 0. The SFF is 76%. According to Figure 7-8. Type A Architecture Requirements IEC 61508, the subsystem qualifies for SIL 2. The overall design is qualified to SIL 1 since lowest subsystem is the limiting factor.

For Type B components, the table in IEC 61508 reduces the SIL levels by one to account for the unknown failure modes of the product. This is shown in Figure 7-9.

Sometimes the hardware fault tolerance is confused with redundancy. They are not necessarily the same thing. Sometimes redundant instruments are used to maintain process operation, not to perform the safety function. In those cases, redundancy is not the same as hardware

Safe Failure Fraction	Hardware Fault Tolerance		
TYPE B	0	1	2
< 60 %	Not Allowed	SIL1	SIL2
60 % < 90 %	SIL1	SIL2	SIL3
90 % < 99 %	SIL2	SIL3	SIL4
> 99 %	SIL3	SIL4	SIL4

Figure 7-9. Type B Architecture Requirements IEC 61508

EXAMPLE 7-9

Problem: Two smart transmitters have been chosen for a SIF design. The logic solver is programmed to trip if either transmitter indicates a dangerous condition (1oo2). The manufacturer's data sheet lists the SFF as 78.4%. To what SIL level is this design qualified per IEC 61508 hardware fault tolerance requirements?

Solution: The design has a hardware fault tolerance of 1 since one instrument can fail and the SIF can still perform the safety function. The SFF is between 60% and 90%, therefore the design qualifies for SIL 2.

---

fault tolerance as defined in IEC 61508 and ANSI/ISA-84.00.01-2004 (IEC 61511 Mod). A chart showing architectures by name and the equivalent hardware fault tolerance is shown in Figure 7-10.

The field chart from ANSI/ISA-84.00.01-2004 (IEC 61511 Mod) can be created from the charts of IEC 61508 by assuming that Type A field components will have a SFF between 0 - 60% and all Type B field components will have a SFF between 60% - 90%. This assumption allowed the chart to be simplified. Unfortunately instruments with a SFF greater than 90% are penalized by the simple chart unless a "prior use" analysis is done. Using the IEC 61508 charts (which are absolutely allowed) for high quality products with a SFF greater than 90% avoids vague prior use documentation.

It can be seen by comparison that if a Type B field component has a SFF of 92% and a hardware fault tolerance of 0, then it meets SIL 2 per Figure 7-8. Using Figure 7-6, the conclusion would be SIL 1 unless a "prior use" justification is documented.

Maximum SIL Allowed: Type B				
	Safe Failure Fraction			
Architecture	0 - <60%	60 - <90%	90 - <99%	99%+
1oo1	Not Allowed	SIL 1	SIL 2	SIL 3
1oo1D	Not Allowed	SIL 1	SIL 2	SIL 3
1oo2	SIL 1	SIL 2	SIL 3	SIL 4
2oo2	Not Allowed	SIL 1	SIL 2	SIL 3
2oo3	SIL 1	SIL 2	SIL 3	SIL 4
2oo2D	Not Allowed	SIL 1	SIL 2	SIL 3
1oo2D	SIL 1	SIL 2	SIL 3	SIL 4
1oo3	SIL 2	SIL 3	SIL 4	SIL 4

Figure 7-10. Architectures versus Architectural Constraints

Low Demand Mode Verification Calculation Using Fault Trees

Fault trees can also be used to calculate PFDavg although one must be careful to calculate PFD as a function of operating time interval and perform the averaging at the top event. Calculating average probabilities for each basic event and combining these probabilities in the fault tree will result in optimistic errors that may result in insufficient safety integrity.

This is because the expected value function is defined using an integral (See Equation 6-1 for PFDavg definition). In any probability combination with an AND gate, probabilities are multiplied. The product of integrals does not equal the integral of products.

Consider a simplex, 1oo1 system. The simplistic approximation equation for PFD is given in Equation F-1. If repair time is short, the first term will be insignificant and the equation can be reduced to:

$$PFD1oo1 = \lambda^{DU} \times TI$$

and the equation for PFDavg given by Equation F-2 can be reduced to:

$$PFDavg1oo1 = \lambda^{DU} \times TI/2$$

If this system were duplicated and configured as a 1oo2 architecture, the fault tree would include an AND gate (Figure F-7). Using a single two input AND gate with PFD as the two inputs, the PFD of the top event would be approximately equal to:

$$PFD1oo2 = (\lambda^{DU} \times TI)^2 \quad \text{(See Equation F-6 for more detail.)}$$

And the PFDavg would equal:

$$PFDavg1oo2 = (\lambda^{DU} \times TI)^2/3.$$ (See Equation F-7 for more detail.)

This approach is correct.

If, however, one used the PFDavg1oo1 as the input to the AND gate, the probability of the top event would be:

$$PFDavg1oo2 = (\lambda^{DU} \times TI)^2/4.$$

This result presents a significant optimistic error and possible SIF designs that do not provide sufficient safety integrity. For three input (or more) AND gates, the results are even more optimistic and therefore should never be used.

Low Demand Mode Verification Calculation Using Markov Models

Markov models may also be used for PFDavg calculations. Like a fault tree the PFD average must be calculated as a function of operating time interval and averaged at the end. This is actually the natural result of solving a Markov model numerically and averaging can be done quite easily with a spreadsheet.

Since Markov models allow precise modeling of repair, this must be done carefully. Details are presented in Appendix D.

Probabilistic Calculation Tools

Given the number of variables needed for accurate low demand probability of failure on demand calculations, a number of engineering tools have been developed. The tools use different methods of calculation and often consider different input variables so care must be taken in the selection of a tool. Some tools are approved by third parties like TUV. Most tools will calculate PFDavg and check the architecture limits based on the SFF. Most will also calculate MTTFS (Mean Time to Fail Spurious). An example output from the SILver^{TM} tool from Exida is shown in Figure 7-11.

Verification Reports

Control system designers must document the results of the conceptual design in sufficient detail for proper implementation. The equipment selected and the redundancy used must be part of the hardware design documentation.

The draft proof test plan and the calculation results must also be part of the package. All input variables used to perform the calculation must be

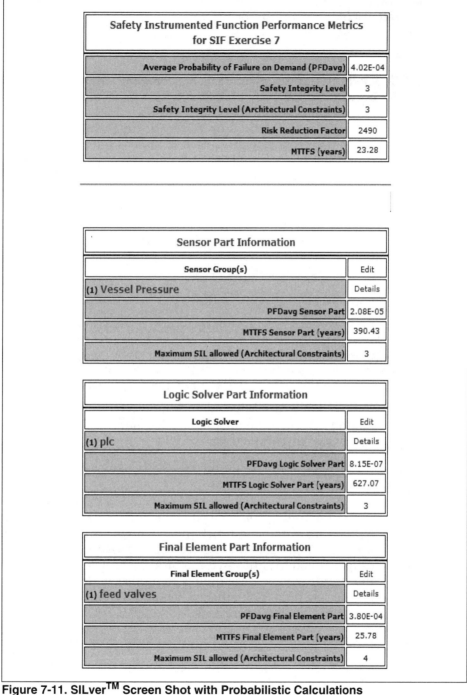

Figure 7-11. SILver™ Screen Shot with Probabilistic Calculations

documented so that the results can be confirmed later if need be. Details should include the tool used with version number, the failure rates used, the failure modes assumed, the proof test frequency, the proof test coverage, repair times assumed, any common cause beta factors and the overall calculation period (the SIF mission time).

Exercises

7-1. Layer of protection analysis has indicated that a demand would occur every 10 years on average for a particular process hazard. Although most automatic diagnostics execute every minute, the worst case time period for automatic diagnostics within the equipment is once per week. A proof test interval of five years is proposed for a manual test and inspection. Would this SIF be classified as high demand or low demand?

7-2. A set of non-redundant (1oo1) safety equipment is used to perform a safety instrumented function. Demands are expected every minute. Diagnostic time is given as thirty seconds. The following failure rate data is obtained when adding the failure rates of the categories of all components:

Lambda SD = 10×10^{-6} failures per hour, Safe Detected

Lambda SU = 5×10^{-6} failures per hour, Safe Undetected

Lambda DD = 8.5×10^{-6} failures per hour, Dangerous Detected

Lambda DU = 0.5×10^{-6} failures per hour, Dangerous Undetected

(NOTE: the terms detected and undetected refer to failures diagnosed by automatic diagnostics, not those detected by the overt false trip of the SIF.)

What is the demand mode? What SIL level is achieved by this design?

7-3. A set of non-redundant (1oo1) safety equipment is used to perform a safety instrumented function. Demands are expected every minute. Diagnostic time is given as one second. The equipment is programmed to shutdown the process on detection of failure and the shutdown response time is 10 milliseconds. The following failure rate data is obtained when adding the failure rates of the categories of all components:

Lambda SD = 10×10^{-6} failures per hour, Safe Detected

Lambda SU = 5×10^{-6} failures per hour, Safe Undetected

Lambda DD = 8.5×10^{-6} failures per hour, Dangerous Detected

Lambda DU = 0.5×10^{-6} failures per hour, Dangerous Undetected

(NOTE: the terms detected and undetected refer to failures diagnosed by automatic diagnostics not those detected by the overt false trip of the SIF.)

What is the demand mode? What SIL level is achieved by this design?

7-4. The total failure rate for a pressure switch is 6×10^{-6} failures per hour. The %Safe is 40%. What is the dangerous failure rate?

7-5. An instrument is to be used in a safety instrumented function in low demand mode. The failure rates are given as:

Lambda SD = 10×10^{-6} failures per hour, Safe Detected

Lambda SU = 5×10^{-6} failures per hour, Safe Undetected

Lambda DD = 8.5×10^{-6} failures per hour, Dangerous Detected

Lambda DU = 0.5×10^{-6} failures per hour, Dangerous Undetected

What is the SFF? If this were a Type B component in a 1oo1 architecture, what SIL level would the design be qualified for per IEC 61508?

7-6. A 2oo3 architecture has what level of hardware fault tolerance?

7-7. A 1oo2D architecture has what level of hardware fault tolerance?

7-8. Two smart transmitters have been chosen for a SIF design. The logic solver is programmed to trip if either transmitter indicates a dangerous condition (1oo2). The manufacturer's data sheet lists the SFF as 92.4%. To what SIL level is this design qualified per IEC 61508? To what SIL level is this design qualified per ANSI/ISA 84.00.01-2004 (IEC 61511) without sufficient "proven in use" evidence? With sufficient "proven in use" evidence?

REFERENCES AND BIBLIOGRAPHY

1. NFPA 85, *Boiler and Combustion Systems Hazards Code*. National Fire Protection Association, 2004.

2. IEC 61508, *Functional Safety of electrical / electronic / programmable electronic safety-related systems*. 2000.

3. ANSI/ISA-84.00.01-2004, *Functional Safety: Safety Instrumented Systems for the Process Industry Sector – Parts 1, 2, and 3 (IEC 61511 Mod)*. 2004.

4. Goble, W.M. *Control Systems Safety Evaluation and Reliability*, Second Edition. ISA, 1998.

5. Goble, W. M. "Accurate Failure Metrics for Mechanical Instruments." *Proceedings of the IEC61508 Conference* (Augsberg, Germany). RWTUV, January 2003.

6. Rosemount, *Quick Installation Guide, Rosemount 3051S Series Pressure Transmitter with HART® Protocol*, 4801 Rev DB, MN: Chanhassen, Rosemount, 2005.

7. *Layer of Protection Analysis*. Center for Chemical Process Safety, American Institute of Chemical Engineers, 2003.

8. Marszal, E. M. and E. W. Scharpf. *Safety Integrity Level Selection: Systematic Methods Including Layer of Protection Analysis*. ISA, 2003.

8
Getting Failure Rate Data

Introduction

When ISA84.01 (Ref. 1) was first released in 1996, several made the comment, "No one has good failure rate data." This led some to believe that the whole idea behind probabilistic failure calculations was impractical.

In the early years of the functional safety standards, industry failure databases could provide failure data information. While this failure data was not product specific or application specific, it helped designers recognize problems in their designs. One such problem was the "weak link" design (Ref. 2). Such a design included a high quality SIL3 safety PLC that was connected to a switch and a solenoid. Many of the engineers thought they had a SIL3 design until they did the safety verification calculations. Such a design may not even meet SIL1 depending on proof test effectiveness and manual proof test time interval!

Even with approximate data, the methods began to show how designers could achieve higher levels of safety while optimizing costs. The safety verification calculations required by the new functional safety standards have shown designers how to design much more balanced designs. The calculations have shown many how to do a better job. But, failure rate and failure mode data for random failures on the chosen equipment is required.

Random Failures versus Systematic Failures

The concept of random failures versus systematic failures was presented in Chapter 3. One must understand the differences in order to understand failure rate data. For safety instrumented function verification calculations, the failure rate data due to random failures during the useful life of a product is required.

The "Well Designed System"

The concept of the "well designed system" was also presented in Chapter 3. A simplistic definition of such a system would be one where all the techniques and measures presented in our functional safety standards to prevent systematic failures are followed. These techniques and measures are planned to significantly reduce the chance of a systematic fault to a tolerable level. Therefore, systematic failure rates caused by human error including failures due to installation errors, failures due to calibration errors and failures due to choosing equipment not suited for purpose are not included in the calculation.

This is not to say that systematic errors cannot happen. It is clearly recognized that these failures do occur and that they do impact safety integrity. One field failure study done by one of the authors traced instrument failure reports to specific end user sites. The results showed that failure rates for the same instrument varied by over an order of magnitude from site to site. There is no doubt that this is significant. But the site specific and even person specific variables preclude an "average" probabilistic approach. That is why it is so important to understand and follow all the procedures, techniques and measures presented in the functional safety standards to avoid and control systematic failures. It is so important to have a "well designed system" for any safety instrumented function.

Industry Failure Databases

Several industry failure databases exist. Analysts gather failure records, make estimates of time in operation and calculate failure rates. The resulting information is published in a book in various forms or provided in a computer database. The main advantage of such documents is that they provide actual field failure based information.

Several problems exist with this method of getting failure rate data. Often needed information about a failure is not collected. This includes total time in operation, failure confirmation, technology class, failure cause and stress conditions. The results are usually a significantly higher failure rate than the number needed for probabilistic SIF verification. This is due to:

1. Lack of distinction between random failures and wear out failures,
2. Lack of distinction between systematic failures and random failures,
3. Merging of technology classes,

4. Incomplete fault isolation, and

5. Other issues.

When total time in operation is not recorded, failures due to wear out cannot be distinguished from random failures during the useful life. If these failures are grouped together, the data analyst cannot distinguish between them and will typically assume that all failures are random. The resulting failure rate number is too high. In addition, the opportunity to establish the useful life period is also lost.

In safety instrumented function verification calculations, the task is to calculate the probability of failure on demand due to random failures. This is done assuming that a preventative maintenance program has been established per the requirements of IEC 61508 (Ref. 3) to replace instruments before the end of their useful life.

When details about failure cause are not collected, failures due to maintenance errors, calibration errors and other systematic faults cannot be distinguished from random failures. The result is a number that can be high.

When failure confirmation is not done, there are times when multiple instruments are replaced during system restoration. When the exact cause of a failure is not identified, multiple "failures" are reported when the maintenance technician replaces several items in an effort to find out which one has actually failed. During a period of unexpected downtime, the emphasis is clearly on system restoration and often time is not allocated for failure identification. This is understandable given many restore situations where there may be a harsh environment, little time and lack of test equipment. The result of recording multiple failures when only one exists is a failure rate that is too high.

In some databases technology classes are mixed. The authors have seen equipment more than fifty years old in operation in industrial processes. Some of this equipment with vacuum tube technology has a significantly different failure rate than solid state integrated circuit technology. When failures from these different technology classes are mixed, the resulting failure rate data is often too high being dominated by the older, less reliable equipment.

In spite of their limitations, industry databases can be extremely valuable especially when no other data source exists. If the failure rate data is too high, the result will be a higher PFH/PFDavg. If this occurs and too much safety integrity is designed into a safety instrumented function, that is tolerable.

Available Databases

One of the most popular failure rate databases is the OREDA database (Ref. 4). OREDA stands for "Offshore Reliability Data." This book presents detailed statistical analysis on many types of process equipment. Many engineers use it as a source of failure rate data to perform safety verification calculations. It is an excellent reference for all who do data analysis.

Other industry failure database sources include:

1. FMD-97: *Failure Mode / Mechanism Distributions*. Reliability Analysis Center, 1997 (Ref. 5)

2. *Guidelines for Process Equipment Reliability Data, with Data Tables*. Center for Chemical Process Safety of AIChE, 1989 (Ref. 6)

3. NPRD-95. *Nonelectronic Parts Reliability Data*. Reliability Analysis Center, 1995 (Ref. 7)

4. IEEE Std. 500. *IEEE Guide To The Collection and Presentation Of Electrical, Electronic, Sensing Component, And Mechanical Equipment Reliability Data For Nuclear-Power Generating Stations*. IEEE, 1984 (Ref. 8)

Many companies have an internal expert who has studied these sources, as well as their own internal failure records, and maintains the company failure rate database. Some use failure data compilations found on the Internet. While the data in industry databases is not product specific or application specific, it does provide useful failure rate information for specific industries (nuclear, offshore, etc.) and a comparison of the data provides information about failure rates versus stress factors.

Failure Mode and Diagnostic Effectiveness Data

Failure rate data alone is not enough to do a good job with probabilistic safety verification. A probability of fail-danger calculation for safety verification purposes requires failure mode data. For each piece of equipment, one must know the failure modes (safe versus dangerous) and the effectiveness of any automatic diagnostics (the diagnostics coverage factor). This information is included only in rough form if at all in industry databases. So many engineers doing safety verification calculations provide an educated and conservative estimate. For most electronic equipment, the safe percentage is set to 50%. Relays have a higher percentage of safe failures with many picking a value of 70% or 80%. Mechanical components like solenoids might be more like 40% safe with many failure modes causing stuck in place failures that end up being dangerous in a safety protection application.

Diagnostic coverage can also be estimated. If "normal" diagnostics are available in a microprocessor based product, diagnostic coverage can be conservatively credited to 50%. Diagnostics for mechanical devices is usually given no credit, 0% detected failures, unless there is some special testing like automatic partial valve stroke testing due to a smart valve positioner.

In spite of their limitations industry databases have served an important purpose. Using a combination of industry databases, company data and experience, the data needed for our safety lifecycle can be estimated. Fortunately other data sources are available also.

Product Specific Failure Data

It is clear that some are uncomfortable with the level of accuracy in the failure data estimated from industry databases and experience. Questions about failure rate versus stress conditions in particular applications come up. Questions about specific products are constantly being asked especially when one must attempt to pick a better product to achieve higher safety.

Fortunately, several instrumentation manufacturers are providing detailed analysis of their products to determine a more accurate set of numbers useful for safety verification purposes. A Failure Modes Effects and Diagnostic Analysis (FMEDA) will provide specific failure rates for each failure mode of an instrumentation product. The percentage of failures that are safe versus dangerous is clear and relatively precise for each specific product. The diagnostic ability of the instrument is precisely measured. Overall, the numbers from such an analysis are indeed product specific and provide a much higher level of accuracy when compared to industry database numbers and experience based estimates.

A FMEDA is done by examining each component in a product. For each failure mode of each component, the random failure rate and effect on the product is recorded. Will this resistor failure cause the product to fail safely, fail dangerously or lose calibration? If the serial communication line from the A/D to the microprocessor gets shorted, how does the product respond? If this spring fractures does that cause a dangerous or a safe failure? The failure rate of each component is entered according to component failure mode and the various categories are added. The end result is a product specific set of failure data that includes failure rates for each failure mode, failure rates that are detected and undetected by diagnostics, safe failure fraction calculations and often an explanation on how to use the numbers to do safety verification calculations.

A FMEDA is sometimes done by the instrument manufacturer but typically done by third party experts. Often a product manufacturer does the work as part of an IEC61508 functional safety certification effort. Many different types of instruments have had this analysis done. A listing of instrumentation assessment including FMEDA analysis is available on www.exida.com/applications/sael/index.asp.

It should be emphasized that a FMEDA provides failure rates, failure modes and diagnostic coverage effectiveness for random hardware failures. If done properly, it does not include failure rates due to "systematic" causes including incorrect installation, inadvertent damage, incorrect calibration or any other human error.

A Comparison of Failure Rates

Failure rates obtained from industry databases, manufacturer FMEDA analysis, manufacturer field failure studies, company failure records or other sources can be compared. The results will be different as described above.

Generally, less specific data turns out to be more conservative and that is appropriate for safety verification purposes following the rule that "the less one knows, the more conservative one must be." Remember that industry databases may include systematic failures, multiple technology classes, wear out failures and possible multiple reports per failure. These issues naturally cause the numbers from such sources to be high.

Table 8-1 shows a comparison of data for a pressure transmitter. The failure rate numbers from the industry database sources are significantly higher than the FMEDA reports.

Comprehensive Failure Data Sources

Recently some analysis organizations have compiled comprehensive failure data source books and computer databases. The information is formatted to give failure rate as a function of failure mode. Often additional information about the product, such as Type A versus Type B, is provided. Some example pages are shown in Figures 8-1, 8-2 and 8-3 (Ref. 9).

The Future of Failure Data

Although product specific FMEDA reports offer superior data sources when compared to industry databases, they still do not account for application specific stress conditions that may affect actual failure rates. Ideally in the future manufacturers will be able to provide not only point estimates of failure rates but perhaps even equations with application specific variables to more precisely calculate the needed numbers. That will happen if there is demand and the needed data is collected.

One effort in the right direction is the PERD (Process Equipment Reliability Database) initiative (Ref. 10) from the Center for Chemical Process Safety (CCPS) of the AIChE (www.aiche.org/ccps/perd/). That

Table 8-1. Failure Rate Data for a Pressure Transmitter

Source	Component	Total Failure Rate (1/hr)	% Safe Failures	Safe Cov. Factor (%)	Dangerous Cov. Factor (%)	Range
CCPS-89	Transmitters - Differential Pressure	1.01E-06	-	-	-	low
	Transmitters - Differential Pressure	6.56E-05	-	-	-	mean
	Transmitters - Differential Pressure	2.54E-04	-	-	-	high
NPRD-95	Transducer, Pressure	8.13E-06	-	-	-	mean
Rosemount	FMEDA, 3051T Pressure Transmitter, exida	4.46E-07	64	100	27.5	FMEDA
Honeywell	FMEDA, ST3000 Pressure Transmitter, exida	4.90E-07	60	100	24.7	FMEDA

Getting Failure Rate Data

exida.com
excellence in dependable automation

TEMPERATURE MEASUREMENTS

EQUIPMENT ITEM	Yokogawa YTA	ITEM NO.	1.6.11

GENERAL INFORMATION

MANUFACTURER	Yokogawa Electric Corporation		
MODEL	YTA		
ANALOG / DIGITAL	Analog		
MEASUREMENT TYPE	Temperature		
ARCHITECTURE TYPE	B	HARDWARE FAULT TOLERANCE	0
DATA SOURCE	FMEDA by exida		
REMARKS	It is assumed that upon detection of Low and High failures of the T/C or RTD the transmitter is programmed to drive the output low.		

FAILURE RATE DATA — PER 10^9 HOURS [FITS]

	TRANSMITTER CONFIGURATIONS			INCLUDING SENSING DEVICES		
	T/C	2-/3-WIRE RTD	4-WIRE RTD	T/C	2-/3-WIRE RTD	4-WIRE RTD
FAIL LOW	351	351	351	5101	1951	2331
FAIL HIGH	27	27	27	27	27	27
FAIL DANGEROUS DETECTED						
FAIL DANGEROUS UNDETECTED	120	120	120	370	520	140
FAIL SAFE DETECTED						
FAIL SAFE UNDETECTED						
FAIL NO EFFECT	169	169	169	169	169	169
SFF [%]	82.0	82.0	82.0	93.5	80.5	94.8

APPLICATION EXAMPLE – INCLUDING 4-WIRE RTD

APPLICATION	PLC DETECTION BEHAVIOR	λ^{SD}	λ^{SU}	λ^{DD}	λ^{DU}	SFF [%]
LOW TRIP	< 4 MA	2331			167	93.7
LOW TRIP	> 20 MA		2331	27	140	94.8
LOW TRIP	< 4 MA & > 20 MA	2331		27	140	94.8
LOW TRIP	-		2331		167	93.7
HIGH TRIP	< 4 MA		27	2331	140	94.8
HIGH TRIP	> 20 MA	27			2471	7.3
HIGH TRIP	< 4 MA & > 20 MA	27		2331	140	94.8
HIGH TRIP	-		27		2471	7.3

SAFETY EQUIPMENT RELIABILITY HANDBOOK © 2003 exida.com L.L.C.

Figure 8-1. SERH FMEDA Based Data Page (Ref. 9) (reprinted with permission of exida)

group has defined data gathering techniques (Ref. 11) and failure taxonomies for various types of process equipment. The important data that must be collected for a failure event has been defined. Operating companies from chemical, petrochemical, industrial gases and other industries become members and are working to set up inspection and failure reporting. They have created data collection software that members use to report field failures to a central database. There is potential that this

Getting Failure Rate Data

EQUIPMENT ITEM	Generic DP / Pressure Switch	**ITEM NO.**	1.4.1

exida.com — Pressure Measurements

GENERAL INFORMATION

MANUFACTURER	Generic equipment
MODEL	~~~
ANALOG / DIGITAL	Digital
MEASUREMENT TYPE	Pressure
ARCHITECTURE TYPE	A
HARDWARE FAULT TOLERANCE	0
DATA SOURCE	exida Comprehensive Analysis
REMARKS	None

FAILURE RATE DATA — Per 10^9 hours [FITs]

	IN CLEAN SERVICE	IMPULSE LINE PLUGGING LIKELY
FAIL LOW		
FAIL HIGH		
FAIL DANGEROUS DETECTED		
FAIL DANGEROUS UNDETECTED	3600	28600
FAIL SAFE DETECTED		
FAIL SAFE UNDETECTED	2400	2400
FAIL NO EFFECT		
SFF [%]	40.0	7.7

Figure 8-2. SERH Generic Failure Data – Switch (Ref. 9) (reprinted with permission of exida)

information could someday become the best possible source of product specific and application specific failure rate and failure mode data. We look forward to better data with more accuracy as we move forward.

126   Getting Failure Rate Data

PRESSURE MEASUREMENTS   exida.com
                                                         excellence in dependable automation

| EQUIPMENT ITEM | Generic DP / Pressure Transmitter | ITEM NO. | 1.4.2 |

GENERAL INFORMATION

MANUFACTURER	Generic equipment
MODEL	~~~
ANALOG / DIGITAL	Analog
MEASUREMENT TYPE	Pressure
ARCHITECTURE TYPE	B
HARDWARE FAULT TOLERANCE	0
DATA SOURCE	exida Comprehensive Analysis
REMARKS	None

FAILURE RATE DATA — Per 10^9 Hours [FITs]

	IN CLEAN SERVICE	IMPULSE LINE PLUGGING LIKELY	
FAIL LOW	750	750	
FAIL HIGH	150	150	
FAIL DANGEROUS DETECTED			
FAIL DANGEROUS UNDETECTED	600	25600	
FAIL SAFE DETECTED			
FAIL SAFE UNDETECTED			
FAIL NO EFFECT			
SFF [%]	60.0	3.4	

APPLICATION EXAMPLE – IN CLEAN SERVICE

APPLICATION	PLC DETECTION BEHAVIOR	λ^{SD}	λ^{SU}	λ^{DD}	λ^{DU}	SFF [%]
LOW TRIP	< 4 mA	750			750	50.0
LOW TRIP	> 20 mA		750	150	600	60.0
LOW TRIP	< 4 mA & > 20 mA	750		150	600	60.0
LOW TRIP	-		750		750	50.0
HIGH TRIP	< 4 mA		150	750	600	60.0
HIGH TRIP	> 20 mA	150			1350	10.0
HIGH TRIP	< 4 mA & > 20 mA	150		750	600	60.0
HIGH TRIP	-		150		1350	10.0

© 2003 exida.com L.L.C.   SAFETY EQUIPMENT RELIABILITY HANDBOOK   55

Figure 8-3. SERH Generic Failure Data – Transmitter (Ref. 9) (reprinted with permission of exida)

Exercises

8-1.   What are the key objectives and differences between FMEA and FMEDA analysis?

8-2.   Explain why the failure rate data for SIS components obtained from field data can be considerably different from that obtained by means of FMEDA analysis for the components.

8-3. A valve manufacturer has issued a certificate for their valve stating that according to IEC 61508 the Mean Time between Failures (MTBF) for the valve was found to be 12000 years. No additional performance data was listed on the certificate. Can this data be used for SIL verification calculations?

8-4. The failure rate for an electronic component is 15.6 FITS. What is the equivalent failure in units of failures per hour?

8-5. A FMEDA report has been issued for a Solenoid valve based on IEC 61508 requirements. What parts of the solenoid valve would have been included in the FMEDA analysis?

8-6. List some of the limitations in using industry database data for SIL verification calculations.

8-7. Does FMEDA based failure rate data include failure rates due to expected human error?

8-8. Based on the information in this Chapter, what is the expected dangerous undetected failure rate for a generic DP/Pressure transmitter?

REFERENCES AND BIBLIOGRAPHY

1. ANSI/ISA S84.01-1996. *Application of Safety Instrumented Systems for the Process Industries.*

2. Gruhn, P. "Safety Systems: Peering Past the Hype." *Proceedings of the 51st Instrumentation Symposium for the Process Industries.* Texas A&M University and ISA, 1996.

3. IEC 61508. *Functional Safety of electrical / electronic / programmable electronic safety-related systems*, 2000.

4. OREDA – 97. *Offshore Reliability Data.* DNV Industry, 1997.

5. *FMD – 97: Failure Mode / Mechanism Distributions.* Reliability Analysis Center, 1997.

6. *Guidelines for Process Equipment Reliability Data.* Center for Chemical Process Safety, American Institute of Chemical Engineers, 1989.

7. NPRD-95. *Nonelectronic Parts Reliability Data.* Reliability Analysis Center, 1995.

8. IEEE Std. 500. *IEEE Guide To The Collection and Presentation Of Electrical, Electronic, Sensing Component, And Mechanical Equipment Reliability Data For Nuclear-Power Generating Stations.* IEEE, 1984.

9. *Safety Equipment Reliability Handbook.* exida, 2003. (available from ISA)

10. Arner, D. C. and W. C. Angstadt. "Where For (sic) Art Thou Failure Rate Data." *ISA Technology Update,* 2001.

11. *Guidelines for Improving Plant Reliability Through Data Collection and Analysis.* Center for Chemical Process Safety, American Institute of Chemical Engineers, 1998.

9

SIS Sensors

Instrument Selection

Sensors in a safety instrumented system measure process variable conditions in order to recognize a potential hazard. Usually these are the same process variables that are used for control. So the first and perhaps most important consideration when selecting sensors for safety applications is that they accurately and reliably measure the process variable. Another key parameter is that any process wetted materials must be compatible with the chemicals of the process. These are two of the key principles required in a "well designed system."

Do not select instruments only because they are safety "certified." Actual performance is the first consideration. This is especially true in level and flow where many different types of measurement technology exist. Of course, if the right technology exists and one manufacturer has done a complete IEC 61508 (Ref. 1) assessment, then that product might have a considerable advantage over another equally able to measure the process.

Many different types of sensors are used in safety instrumented functions. An informal survey was done as part of a market study (Ref. 2). Results regarding sensor usage are shown in Figure 9-1.

Diagnostic Annunciation

Sensors designed for safety instrumented system applications typically have excellent built-in automatic diagnostics. That attribute is one of the main advantages of sensors designed per IEC 61508. However, it must be remembered that automatic diagnostics must be annunciated so that a repair team can quickly restore correct operation. Without annunciation and effective repair, the diagnostics do little good.

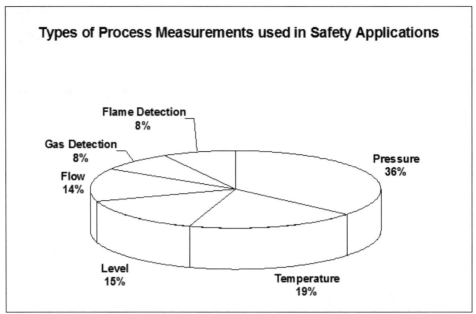

Figure 9-1. Informal Survey of Process Variables used in Safety Applications

Many transmitters that use a 4–20 milliamp analog current to signal the process variable also use the current level to signal an internal fault detected by the automatic diagnostics in the transmitter. Although the current levels do vary, one common set of values based on the NAMUR NE-43 standard is shown in Figure 9-2.

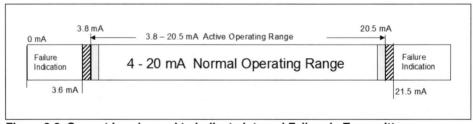

Figure 9-2. Current Levels used to Indicate Internal Failure in Transmitters

Other sensors, particularly fire and gas sensors with external power commonly use a one milliamp or a two milliamp current level to indicate an internally detected fault. For all these cases, the logic solver must be able to read those current levels and be programmed to interpret those levels as a diagnostic fault. This should be done with a filter or timer to insure that a transitioning current level does not cause a false trip. When diagnostic annunciation is done correctly, the probabilistic modeling can give credit for the automatic detection. Safety is improved and false trips can be avoided.

Probabilistic Modeling of Sensors

Failure rate data for sensors is often first generated using functional failure modes as it is often not possible to know which failures are safe versus dangerous at the product level. As an example, consider the pressure transmitter failure data in terms of functional failure modes per the data in Table 9-1.

Table 9-1. Pressure Transmitter Functional Failure Data

Pressure Transmitter Failure Mode	Failure Rate	
Fail output high > 21.5 mA	0.00131	Failures per year
Fail output low < 3.6 mA	0.00219	Failures per year
Fail output frozen	0.00175	Failures per year
Fail output drifting	0.00350	Failures per year
Failure detected	0.00438	Failures per year
Failure of diagnostics	0.00013	Failures per year

At the safety instrumented function level, the functional failure modes of a pressure transmitter might cause a false trip, might cause a dangerous failure or may generate a diagnostic alarm in the safety PLC. The application must be considered when classifying these failure rates into categories such as safe, dangerous, no effect and so forth.

Table 9-2. Failure Rate Categories

Failure Mode	Failure Rate	DD	DU	AU
Fail output high > 21.5 mA	0.00131	0.00131		
Fail output low < 3.6 mA	0.00219	0.00219		
Fail output frozen	0.00175		0.00175	
Fail output drifting	0.00350		0.00350	
Failure detected	0.00438	0.00438		
Failure of diagnostics	0.00013			0.00013
	Totals f/year	0.00788	0.00526	0.00013
	Totals f/hour	0.0000009	0.0000006	0.000000015

EXAMPLE 9-1

Problem: A single pressure transmitter (1oo1) is being used in a safety instrumented function to initiate a trip when pressure goes above 80% of scale (16 mA). This transmitter is connected to a safety PLC programmed to detect under-range (< 3.6 mA) and over-range (> 21.5 mA) and send a diagnostic alarm. The safety PLC has a filter on the analog input to prevent any spurious trip when the current makes its transition from active value to failure condition. The pressure transmitter is programmed to send its output low on detection of a failure. How are the failures of Table 9-1 classified?

Solution:

1. When the transmitter fails with its output saturated over-range (> 21.5 mA), the safety PLC will automatically detect this as a failure and send an alarm. No false trip of the safety instrumented function will occur as the PLC is programmed to recognize this not as a trip but as a diagnostic fault in the transmitter. However, the transmitter is not capable of responding to a demand during this time so the failure should be classified as **Dangerous Detected**.

2. When the transmitter fails with its output saturated under-range (< 3.6 mA), the safety PLC will automatically detect this as a failure and send an alarm. No false trip will occur. As above, the transmitter is not capable of responding to a demand during this time so the failure should be classified as **Dangerous Detected**.

3. When the transmitter fails with its output frozen it is not capable of responding to a demand. Since this is not detected by any internal transmitter diagnostics and the current will be in the active range, it is classified as **Dangerous Undetected**.

4. When the transmitter fails with its output drifting, the drift may go in either direction. So this failure mode could be classified as either safe or dangerous. However, since the drift is usually unpredictable it must be classified in worst-case mode. Since this is not detected by any internal transmitter diagnostics and the current will be in the active range, it is classified as **Dangerous Undetected**.

5. Internal failures that are detected by automatic diagnostics within the transmitter will send the current level to the value programmed in the transmitter. In this case the problem statement tells us that the current level will go low (< 3.6 mA). As long as the current level goes out of range the actual level does not matter as the safety PLC will detect either high or low and send an alarm to the repair team. However, the transmitter is not capable of responding to a demand during this time so the failure should be classified as **Dangerous Detected**.

6. Failures within the transmitter that cause loss of diagnostics do not affect the safety functionality so could not be classified as either safe or dangerous. The effect of these failures on safety integrity is likely small but these failures could be modeled for more accuracy (Ref. 3). These failures will be identified as **Annunciation Undetected**.

EXAMPLE 9-2

Problem: How would the transmitter of Example 9-1 be modeled in an application with a test interval of five years assuming full inspection with 100% effectiveness and restore time when the failure is detected of 48 hours? What is the PFDavg?

Solution: The failure rates must added as shown in Table 9-2.

The simplest solution would be to use the equations from Appendix F. Using Equation F-2 (remember that the failure rates are in units of failures per year):

$$PFDavg_{1oo1} = \lambda^{DD} \times RT + \lambda^{DU} \times TI/2$$

$$= 0.00788 \times 48/8760 + 0.00526 \times 5/2$$

$$= 0.0000432 + 0.01314$$

$$= 0.01318$$

As an alternative solution, a more detailed Markov model could be created as shown in Figure 9-3.

The P matrix for this model is shown in Figure 9-4.

When this model is solved used numerical techniques and the time dependent PFD values averaged, the result is a PFDavg = 0.013067. The difference between this answer and the previous answer of 0.01318 represents the approximation of the simplified equations.

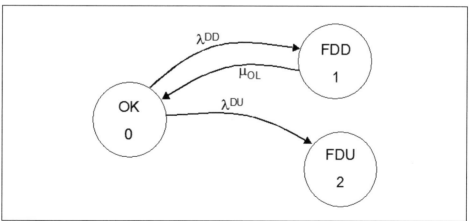

Figure 9-3. Markov Model for Transmitter

The Markov model of Figure 9-3 did not account for the annunciation failures. The annunciation failures could be accounted for using a more detailed Markov model as shown in Figure 9-5.

$$\begin{array}{c} & \begin{array}{ccc} 0 & \quad\quad 1 & \quad\quad 2 \end{array} \\ \begin{array}{c} 0 \\ 1 \\ 2 \end{array} & \left[\begin{array}{ccc} 1 - (\lambda^{DD} + \lambda^{DU}) & \lambda^{DD} & \lambda^{DU} \\ \mu_{OL} & 0 & 1 - (\mu_{OL}) \\ 0 & 0 & 1 \end{array} \right] \end{array}$$

Figure 9-4. P Matrix for Markov Model of Figure 9-3

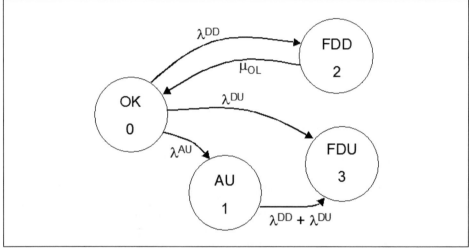

Figure 9-5. Markov Model of Transmitter Example

The P matrix for this model is shown in Figure 9-6.

$$\begin{array}{c} & \begin{array}{cccc} 0 & \quad 1 & \quad 2 & \quad 3 \end{array} \\ \begin{array}{c} 0 \\ 1 \\ 2 \\ 3 \end{array} & \left[\begin{array}{cccc} 1 - (\lambda^{DD} + \lambda^{DU} + \lambda^{AU}) & \lambda^{AU} & \lambda^{DD} & \lambda^{DU} \\ 0 & 1 - (\lambda^{DD} + \lambda^{DU}) & 0 & \lambda^{DD} + \lambda^{DU} \\ \mu_{OL} & 0 & 1 - (\mu_{OL}) & 0 \\ 0 & 0 & 0 & 1 \end{array} \right] \end{array}$$

Figure 9-6. P Matrix for Figure 9-5

When the failure rate numbers are substituted into the matrix, the result is shown in Figure 9-7.

P	0	1	2	3
0	0.999998485	0.000000015	0.0000009	0.0000006
1	0	0.9999985	0	0.0000015
2	0.020833333	0	0.979166667	0
3	0	0	0	1

Figure 9-7. Numeric Values for P Matrix

When this matrix is solved numerically and the time dependent results are averaged, the PFDavg = 0.013072. This answer is slightly higher than the previous simpler model as would be expected. In this situation with the annunciation failure rate low and the diagnostic capability medium, it is probably not worth taking the time to model annunciation failures.

However, if the diagnostic coverage were high, the annunciation failures will have a greater impact. Consider the failure rates from Table 9-3. These values were taken from the FMEDA report for the Rosemount 3051S SIS transmitter (Ref. 4) and converted to units of failures per year.

Table 9-3. Failure Rate Numbers for 3051S SIS

Pressure Transmitter Failure Mode	Failure Rate	
Fail output high > 21.5 mA	0.00054	Failures per year
Fail output low < 3.6 mA	0.00243	Failures per year
Fail dangerous undetected	0.00064	Failures per year
Failure detected	0.00438	Failures per year
Failure of diagnostics	0.00034	Failures per year

Using these values, the PFDavg calculated by the simplified equation is 0.001639. The Markov model without the annunciation failures provides a result of 0.001637. The Markov model with the annunciation failures provides a result of 0.001647.

Pressure

The most common measurement in safety instrumented system applications is pressure. Fortunately, the instrumentation products available for this application are quite mature and highly advanced. There are well proven products available that are also IEC 61508 assessed available from more than one vendor.

The Rosemount 3051S SIS (Figure 9-8) received its IEC 61508 assessment rating in 2004. As one of most popular transmitters in the process control market, the 3051S has accumulated substantial proven in use hours during the years it has been on the market. The product Safety Manual provides failure rates, failure modes, suggested proof test procedures, proof test coverage estimates, a common cause beta factor estimate range and all settings that should be considered in a safety application.

Figure 9-8. Rosemount 3051S SIS Transmitter (used with permission of Rosemount)

The 3051S SIS has a 61508 assessment certificate states that the product can be used in SIL 2 applications as a single transmitter and SIL 3 applications if more than one transmitter is used in an identical redundant (hardware fault tolerance > 0) architecture. This helps point out the differences between random and systematic failures. The design process used to create the transmitter and its software met the more rigorous criteria of SIL 3. The chance of a systematic fault is lower.

Many engineers are very concerned about "common cause" failures in redundant architectures. Some engineers will not use two identical products in a redundancy scheme for safety. The reasoning is that two identical products are far more likely to fail due to common cause. This renders the redundancy less effective. Studies have backed up this position (Ref. 5, 6, 7, and 8).

However, it was recognized that many common cause failures of identical redundant products were caused by systematic design faults. Therefore some products like the 3051S are assessed per IEC 61508 to a higher SIL level for the product design process to allow for use of identical redundant designs.

The advantages of identical redundancy include fewer spare parts, better knowledge of maintenance procedures (less chance of maintenance error) and faster repair times.

The Yokogawa EJX is another popular transmitter (Figure 9-9) that has received full IEC 61508 assessment during 2004. It has been assessed to SIL 2 using a single transmitter and SIL 3 for identical fault tolerant architectures. As before, this tells us that the product design and test process met the more strenuous requirements of SIL 3 so that identical redundancy designs are acceptable.

Figure 9-9. Yokogawa EJX Transmitter (used with permission of Yokogawa)

A partial list of products with some level of assessment is shown in Table 9-4.

Table 9-4. Partial List of Pressure Transmitter Safety Assessments

Manufacturer	Model	Assessment Level	Assessor
ABB	Safety 600T	61508 Assessment	TUV Sud
ABB	2600T (268 model)	61508 Assessment	TUV Sud
ABB	2600T (265, 267, 269 models)	exida Proven in Use	exida
Endress+Hauser	Cerebar S	61508 Assessment	exida/TUV Sud
Endress+Hauser	Deltabar S	61508 Assessment	exida/TUV Sud
Honeywell	ST3000	FMEDA	exida
Prime Measurement	Model 345	61508 Assessment	TUV Sud
Rosemount Inc.	3051T	FMEDA	exida
Rosemount Inc.	3051 S SIS	61508 Assessment	exida/RWTUV
Rosemount Inc.	3051C	FMEDA	exida
Rosemount Inc.	3051S	FMEDA	exida
Siemens AG	SITRANS P	FMEDA	TUV Sud
Smar	LD290, LD291, LD301	FMEDA	exida
SOR	SGT	FMEDA	FM
Yokogawa Electric Corporation	EJA	FMEDA	exida
Yokogawa Electric Corporation	EJX	61508 Assessment	exida/RWTUV
Yokogawa Electric Corporation	UniDelta Mark II	FMEDA	exida

Temperature

Temperature is another common measurement in safety instrumented system applications. Instrumentation products available for temperature measurement are also quite mature and highly advanced. A number of well proven products are available and some of these are also IEC 61508 assessed.

The Rosemount 3144P SIS (Figure 9-10) received its IEC 61508 assessment rating in 2004. It is capable of using two temperature sensing elements and can be set up to provide comparison diagnostics within the transmitter. This feature provides a high diagnostic coverage on the sensing elements. The Rosemount 3144P is rated for SIL2 in single applications and SIL3 in a fault tolerant architecture.

138   SIS Sensors

Figure 9-10. Rosemount 3144P Temperature Transmitter (used with permission of Rosemount)

A partial list of products that have received some level of assessment is shown in Table 9-5.

Table 9-5. Partial List of Temperature Transmitters with Assessment

Manufacturer	Model	Assessment Level	Assessor
ABB	TH02, TH102, TH202	exida Proven In Use	exida
Endress+Hauser	TMT 122 / 182	exida Proven In Use	exida
Endress+Hauser	TMT 162	FMEDA	exida
Honeywell	STT250	FMEDA	exida
Moore Industries Inc.	TRY / TRY DIN	FMEDA	exida
Rosemount Inc.	3144P	61508 Assessment	exida/RWTUV
Rosemount Inc.	644	FMEDA	exida
WIKA	T32	FMEDA	exida
Yokogawa Electric Corporation	YTA	FMEDA	exida

Level

Level is common in many safety instrumented applications as well. It is used in separation units to prevent high pressure "blow-by" and is common in tank farms. There are a number of different technologies used to measure level.

Sensors are available as either "level transmitters" that send a proportional analog signal or "level switches" that send a Boolean value. Both types of products are available with a two wire 4–20 mA interface. Level switches are also available with contact switch output.

Figure 9-11 shows a picture of the Endress + Houser Liquiphant Fail Safe level switch. This product was the first sensor of any type to receive a safety assessment in 1998. It currently has received a full IEC 61508 assessment.

Figure 9-11. Endress + Houser Liquiphant Fail Safe Level Switch (used with permission of E+H)

A partial list of level transmitters with some assessment is shown in Table 9-6.

Table 9-6. Partial List of Level Transmitters with Assessment

Manufacturer	Model	Assessment Level	Assessor
Endress+Hauser	Micropilot M	exida Proven In Use	exida
Endress+Hauser	Levelflex M	exida Proven In Use	exida
K-TEK Corporation	MT2000	FMEDA	exida
K-TEK Corporation	AT500 / AT600	FMEDA	exida
K-TEK Corporation	AT100 / AT200	FMEDA	exida
Magnetrol	Eclipse Model 708	FMEDA	exida
Magnetrol	Eclipse Model 705	FMEDA	exida
VEGA Grieshaber	VEGAPULS 4x / 5x	exida Proven In Use	exida

A partial list of level switches with some assessment is shown in Table 9-7.

Table 9-7. Partial List of Level Switches with Assessment

Manufacturer	Model	Assessment Level	Assessor
Ametek Drexelbrook	Intellipoint RF	FMEDA	exida
Endress+Hauser	Liquiphant Fail Safe	61508 assessment	TUV Sud
Endress+Hauser	Liquiphant M/S	exida Proven In Use	exida
Endress+Hauser	FTL325/375P+FEL57	61508 assessment	TUV Sud
Magnetrol	915P/915W	FMEDA	exida
VEGA Grieshaber	VEGAVIB 60	FMEDA	exida
VEGA Grieshaber	VEGASWING 61/63	61508 assessment	exida

Flow

Flow measurement is important to many safety instrumented functions. As experienced instrumentation engineers will advise, sensors used in safety instrumented function level applications must be very carefully applied. There are many different technologies available with different capabilities and features. The safety instrumented function designer must choose the instrument best capable of accurately measuring the flow. Fortunately there are many products available with safety assessment completed.

The Micro-Motion 1700/2700 Coriolis flowmeter (Figure 9-12) received its IEC 61508 assessment in 2005. The basic technology is frequency based providing high inherent safety as most internal failures will result in lack of frequency and are therefore detectable by the transmitter.

Figure 9-12. Micro-Motion 1700/ 2700 Flow Transmitter (used with permission of Micro-Motion)

A partial list of flow transmitters with some safety assessment is shown in Table 9-8.

Table 9-8. Partial List of Flow Transmitters with Assessment

Manufacturer	Model	Assessment Level	Assessor
ABB	FCM2000 - Coriolis Mass Flowmeter	FMEDA	exida
Endress+Hauser	PROMASS 80/83 Coriolis Mass Flowmeter	exida Proven In Use	exida
Micro-Motion	Model 2700 Coriolis Multivariable Flow	61508 assessment	exida/RWTUV
Micro-Motion	Model 1700 Coriolis Multivariable Flow	61508 assessment	exida/RWTUV
Rosemount Inc.	8800C Vortex Flow	FMEDA	exida
Rosemount Inc.	8712D	FMEDA	exida
Rosemount Inc.	8732 Magnetic Flow	FMEDA	exida
Yokogawa Electric Corporation	DY/DYA Vortex Flow	FMEDA	exida

Gas/Flame Detectors

Although some consider fire and gas systems to be outside the scope of a safety instrumented system, many others classify these functions as safety instrumented functions. The criteria is based on needed risk reduction. When consequence or likelihood reduction is achieved by these functions and the risk reduction needed is greater than 10, ANSI/ISA-84.00.01 (IEC 61511 Mod) (Ref. 9) requires the function be classified as a SIF.

Most of the products in this category have been designed as automatic protection devices and many of the products available on the market have received some level of IEC 61508 assessment.

Figure 9-13 shows a picture of the Det-Tronics Pointwatch Eclipse IR Gas Detector. This product received full IEC 61508 assessment to SIL 2 in 2005.

Figure 9-13. Det-Tronics Pointwatch Eclipse IR Gas Detector (used with permission of Detronics)

The Det-Tronics X3301 Flame Detector received IEC 61508 assessment to SIL 2 in 2005. It is shown in Figure 9-14.

A partial list of gas detectors with safety assessment is shown in Table 9-9 and a partial list of flame detectors is shown in Table 9-10.

Figure 9-14. Det-Tronics X3301 Flame Detector (used with permission of Det-Tronics)

Table 9-9. Partial List of Gas Detectors with Assessment

Manufacturer	Model	Assessment Level	Assessor
Det-Tronics	Pointwatch Eclipse IR	61508 assessment	exida/RWTUV
Dräger Safety	Polytron 2IR - Type 334	exida Proven In Use	exida
Dräger Safety	Polytron 7000	FMEDA	exida
Dräger PLMS ltd.	Polytron Pulsar	FMEDA	exida
Zellweger	APEX	FMEDA	exida

Table 9-10. Partial List of Flame Detectors with Assessment

Manufacturer	Model	Assessment Level	Assessor
Det-Tronics	X3301 Multispectrum IR	61508 assessment	exida/RWTUV

Burner Flame Detectors

Flame detectors designed for burner protection are quite different than flame detectors designed for fire and gas systems. Burner flame detectors monitor the flame inside a combustion chamber and look for lack of flame. The dangerous condition is loss of flame.

Conversely, in detectors designed for fire and gas mitigation systems; presence of flame is the dangerous condition. Since the dangerous situation is reversed for these two applications, the overall design of the sensor is different especially the fail-safe characteristics.

A partial list of burner flame detectors that has had some level of safety assessment is shown in Table 9-11.

Table 9-11. Partial List of Burner Flame Detectors with Assessment

Manufacturer	Model	Assessment Level	Assessor
Fireye	Phoenix 85UVF	FMEDA	exida
Fireye	Insight 951R/95UV/95DS	FMEDA	exida

It should be noted that several manufacturers of burner flame detectors have had their product assessed per EN298, a European standard. This standard is similar to IEC 61508 but does not require publication of failure rates and failure modes nor does it require publication of safety manual. Instruments assessed to this standard should provide a high level of safety however.

Miscellaneous

The authors have seen many other types of sensors used in safety instrumented functions. These include:
- AC electric current sensors
- Mass spectrometers
- Oxygen analyzers
- Electric voltage sensors
- Millivolt sensors
- Etc.

The great variety reflects the entire range of instrumentation used in process control.

Exercises

9-1. What is the most important criteria to use when selecting sensors for safety instrumented function applications?

   a. ability to accurately measure the process variable

   b. IEC 61508 assessment

   c. purchase cost

   d. manufacturers quality system rating

9-2. Failure rate and failure mode data is available for which sensors:

   a. pressure and temperature only

   b. pressure, temperature, flow and level

   c. all sensors.

9-3. What type of flame detector would be chosen for a burner management system?

9-4. When would a fire and gas function be classified as a safety instrumented function?

9-5. When would a burner management system function be classified as a safety instrumented function?

REFERENCES AND BIBLIOGRAPHY

1. IEC 61508, *Functional Safety of electrical / electronic / programmable electronic safety-related systems*, 2000.

2. *Sensor Market Study.* Exida, 2003.

3. Goble, W. M. and J. V. Bukowski. "Extending IEC61508 Reliability Evaluation Techniques to Include Common Circuit Designs Used in Industrial Safety Systems." *Proceedings of the Annual Reliability and Maintainability Symposium.* IEEE, 1997.

4. *Failure Modes, Effects and Diagnostic Analysis*, ROSO2/11-07 ROO1, PA: Sellersville, exida, Feb. 2004.

5. Hokstad, P. and L. Bodesberg. "Reliability Model for Computerized Safety Systems." *Proceedings of the Annual Reliability and Maintainability Symposium.* IEEE, 1989.

6. Rutledge, P.J. and A. Mosleh. "Dependent-Failures in Spacecraft: Root Causes, Coupling Factors, Defenses, and Design Implications." *Proceedings of the Annual Reliability and Maintainability Symposium.* IEEE, 1995.

7. Gole, W. M., Bukowski, J. V. and Brombacher, A. C., "How Common Cause Ruins the Safety Rating of a Fault Tolerant PES," *Proceedings of the ISA Spring Symposium - Cleveland.* NC: Research Triangle Park, ISA, 1996.

8. Bukowski, J. V., and A. Lele. "The Case for Architecture-Specific Common Cause Failure Rates and How They Affect System Performance." *1997 Proceedings of the Annual Reliability and Maintainability Symposium.* IEEE, 1997.

9. ANSI/ISA-84.00.01-2004, *Functional Safety: Safety Instrumented Systems for the Process Industry Sector – Parts 1, 2, and 3 (IEC 61511 Mod).* 2004.

10
Logic Solvers

Introduction

In a safety instrumented function the logic solver provides the intelligence to perform any needed logic as well as many other potential functions such as filtering, averaging, comparison and calculations. There are several types of technologies that have been used including relays, pneumatic logic, solid state logic, intrinsically safe solid state logic, programmable logic controllers, distributed control systems and purpose built safety programmable logic controllers.

Relays/Pneumatic Logic

Electromechanical relays and pneumatic logic have been used for many decades to perform logical functions in automated safety protection systems. These circuits can be wired (or piped) to perform AND, OR and even 2oo3 voting functions. Time delay relays provide filtering functionality. "Trip Amplifiers" are also available that convert analog inputs into a digital trip signal for use with relay logic. Generally such circuits have a bias toward a fail de-energized mode.

Special relays called "Safety Relays" are even constructed with multiple springs, protection against welded contacts, positive guided contacts, mechanically connected multiple contacts for monitoring and sealed construction. These components are strongly biased to the de-energized failure mode and generally have a safety certification. With such components the designer can create a safety instrumented function with highly predictable "fail-safe" characteristics; a primary advantage of the relay logic approach. A safety relay is shown in Figure 10-1.

Relays are considered a good choice for the logic solver when the I/O count is low (less than 6) and future logic changes are highly unlikely.

Figure 10-1. Safety Relays

Given the availability of a small, inexpensive safety PLC, relays are typically not the optimal choice of logic solver in many designs.

Solid State / Intrinsically Safe Solid State

As the logic got more complicated, systems expanded to include large panels packed with relays and timers. It was perhaps natural for some engineers to convert this logic to a new "solid state" design when these solid state logic components became available in the late 1960s. Unfortunately there was little consideration of the component failure modes of these designs.

When it was clearly recognized that the failure mode of this logic was not predictable, special "intrinsically safe" logic systems were created. Using dynamic signals that depended on magnetic (Figure 10-2) or capacitive coupling of energy, these logic modules were carefully designed to fail de-energized.

These designs had predictable failure modes and provided good space density and weight compared to safety relays. But like safety relays, these systems required careful design procedures to reduce the chance of an implementation error. These systems could be hard to maintain especially in a situation where logic changes were necessary.

Programmable Logic Controllers

Programmable logic controllers (PLC) often seem a logical choice for safety instrumented system logic solvers. After all, the PLC was designed

Figure 10-2. Magnetic Logic Module

to replace relays and relays are appropriate for safety. The additional advantages of the PLC include ease of engineering with graphical displays of the logic, program error checking utilities, quick documentation capability, built-in testing tools, communications to display devices and internal diagnostics.

In spite of these advantages, the failure rates and failure modes of typical PLC designs will allow these devices to reach only SIL 1 (see Example 10-2) if that. Of course, the higher quality products are more likely to have safety class diagnostics. Most manufacturers of these product lines are getting IEC 61508 (Ref. 1) assessments after which the product is known as a "safety PLC."

Safety Programmable Logic Controllers

In order to assure that the programmable logic controllers were designed to strongly fail in a safe manner specific design rules were created (Ref. 2, 3, 4, and 5) and published by regulatory bodies and technical agencies. These documents defined what became known as a "safety PLC," a product specifically designed to accomplish a key objective, fail only in a predictable, safe way. The principle applies to both random hardware failures and systematic (software) design failures; the rules are meant to keep both types of faults to acceptable levels according to the safety integrity level.

To protect against random hardware failures, a safety PLC will emphasize internal automatic diagnostics, a combination of hardware and software that allows the machine to detect improper operation within itself. Special electronic circuitry, careful diagnostic software analysis, and full fault injection testing of the complete design insure that a safety PLC is capable of detecting potentially dangerous internal component failures. A Failure Modes, Effects and Diagnostic Analysis (FMEDA) is conducted on the design, indicating how each component in the system fails and how the system detects the failure. The level of diagnostics matches the SIL level, better diagnostics or more hardware fault tolerance is required for SIL 3.

To protect against systematic faults, a safety PLC will have software that uses a number of special techniques to ensure software reliability. A safety PLC will have extra security on any reading and writing via a digital communications port. The software is analyzed to avoid complexity. Extensive analysis called "Software Criticality Analysis" and testing carefully examines all software tasks and operating systems for task interaction. This testing includes real-time interaction, such as multi-tasking (when used) and interrupts. Special diagnostics called "program flow control" and "data verification" are required. Program flow checking ensures that essential functions execute in the correct sequence. Data verification stores all critical data redundantly in memory and checks validity before use. The analysis is more detailed for SIL 3 than for SIL 2.

To verify data integrity checking, a series of "software fault injection" tests must be run. The programs are deliberately corrupted to ensure that the PLC responds in a predictable, safe manner. The software design and testing is fully documented so that third-party inspectors can understand PLC operation. While some software development teams do not utilize these techniques, they are precisely how the most insidious software design faults are uncovered. The level of testing is more extensive for SIL3 than for SIL 2. That is why multiple SIL 2 machines do not provide the level of safety as one SIL 3 machine.

Many were introduced to the concept of a safety PLC in the mid-1980s when Honeywell began marketing a controller manufactured by Triconex for safety critical applications (Figure 10-3). Since that time many PLC manufacturers have had their products assesed per IEC 61508.

Table 10-1 shows a partial list of safety PLC products.

Logic Solvers 149

Figure 10-3. Triconex Safety PLC (used with permission of Triconex)

Table 10-1. Partial List of Safety PLC Products

Manufacturer	Model	Description	Assessment Type	Assessment Agency
ABB	Safeguard 400	Safety PLC	61508 Certification	TÜV Sud
Det-tronics	Eagle Quantum	Fire and Gas Controller / PLC	FMEDA	exida
Emerson	DeltaV SIS	Safety PLC	61508 Certification	exida/RWTÜV
HIMA	A1/A1 dig	Safety PLC	61508 Certification	TÜV Rheinland
HIMA	H41/H51q	Safety PLC	61508 Certification	TÜV Sud
Honeywell	FSC	Safety PLC	61508 Certification	TÜV Sud
ICS Triplex	Trusted	Safety PLC	61508 Certification	TÜV Rheinland
Siemens AG	S7-400F, S7-400F/H	Safety PLC	61508 Certification	TÜV Sud
Siemens AG	S7-300 F	Safety PLC	61508 Certification	TÜV Sud
Siemens AG	QUADLOG	Safety PLC	61508 Certification	TÜV Sud
Triconex	Trident	Safety PLC	61508 Certification	TÜV Rheinland
Triconex	Tricon V9	Safety PLC	61508 Certification	TÜV Rheinland
Yokogawa Electric Corporation	Prosafe PLC	Safety PLC	61508 Certification	TÜV Sud
Zellweger Analytics	System 57	Fire and Gas System	FMEDA	exida

Probabilistic Modeling of the PLC

When doing probabilistic modeling for safety verification purposes for a PLC, the modeler must identify what portion of the PLC is being used for the each safety instrumented function. Consider the functional subsystems of a typical single channel (1oo1) multi-module PLC (Figure 10-4).

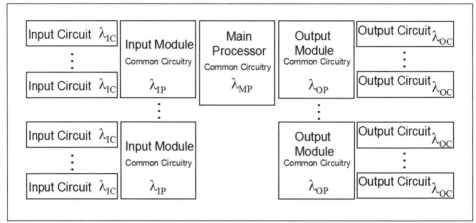

Figure 10-4. Major Subsystems in Multi-module PLC

Each "input circuit" contains the electronics required to read one sensor input. The "input module" subsystem includes all the electronics common to all input channels on a module. The "main processor" encompasses all components common to any PLC function. The "output module" subsystem contains all components common to the output channels on one module. The "output circuit" consists of the components needed to interface to one final elements device.

> **EXAMPLE 10-1**
>
> **Problem:** A safety instrumented function uses three analog transmitters as process sensors, a safety PLC and two remote actuated valves interfaced to the PLC with 24 VDC three way solenoids. What portion of the safety PLC is used in this safety instrumented function?
>
> **Solution:** Three analog input channels are needed as well as all common circuitry in an analog input module. All common circuitry in the PLC is required (Main Processor). The common circuitry for one digital output module is required and two digital output circuits are required.

Probabilistic Modeling of an Advanced Safety PLC

There is a significant difference between probabilistic modeling for a PLC with a simple architecture (1oo1, 1oo2, etc.) and the higher end complex architectures (1oo1D, 2oo2D, 1oo2D [Ref. 6], 2oo4D, 2oo3, etc.), where Markov modeling is the best way to account for all the intermediate degradation states and internal diagnostics. This can be seen to a certain extent in Appendix F and ISA-TR84.00.02, Part 5 (Ref. 7).
There are significant differences in performance that can only be shown by subtle changes in the model (Ref. 8). Simplistic models that do not account for all variables would not be accurate and thus provide the wrong calculation for sophisticated technology. This is the reason why some manufacturers, like Triconex, provide TÜV approved Markov modeling performance calculation sheets for any specific I/O configuration and some TÜV approved safety verification tools, like the exida SILver^{TM} tool, use these models supplied by the manufacturer.

Failure Rate Data for PLC

IEC 61508 requires that a manufacturer provide failure rate and failure mode data so that safety instrumented function designers can perform probabilistic verification of their design. Manufacturers who meet all requirements of the standard will provide a full set of data similar to the example in Table 10-2.

Table 10-2. Failure Rate Data Set for Generic PLC (from exida, reprinted with permission)

LOGIC SOLVERS

EQUIPMENT ITEM	General Purpose PLC		ITEM NO	3.1

GENERAL INFORMATION

MANUFACTURER	Generic equipment		
MODEL	~~~		
LOGIC SOLVER TYPE	PLC	CERTIFIED FOR USE UP TO SIL	N/A
CONFIGURATION	1oo1	BETA FACTOR [%]	N/A
ARCHITECTURE TYPE	B	HARDWARE FAULT TOLERANCE	0
DATA SOURCE	exida Comprehensive Analysis		
REMARKS	None		

FAILURE RATE DATA — PER 10⁹ HOURS [FITS]

	MODEL #	λ^{SD}	λ^{SU}	λ^{DD}	λ^{DU}
MAIN PROCESSOR	N/A	4500	500	3500	1500
POWER SUPPLY	N/A	4513	238	238	13
ANALOG IN MODULE	16 channel	850	150	750	250
ANALOG IN CHANNEL	N/A	25	25	13	38
DIGITAL IN MODULE (1)	16 channel	425	75	375	125
DIGITAL IN CHANNEL (1)	N/A	50	50	25	75
DIGITAL IN MODULE (2)	16 channel	425	75	375	125
DIGITAL IN CHANNEL (2)	N/A	50	50	25	75
ANALOG OUT MODULE	16 channel	850	150	750	250
ANALOG OUT CHANNEL	N/A	125	125	63	188
DIGITAL OUT LOW MODULE	16 channel	425	75	375	125
DIGITAL OUT LOW CHANNEL	N/A	50	50	25	75
DIGITAL OUT HIGH MODULE	16 channel	425	75	375	125
DIGITAL OUT HIGH CHANNEL	N/A	100	100	50	150

APPLICATION EXAMPLE – ARCHITECTURAL CONSTRAINTS

PLC CONFIGURATION	MODULE SFF [%]	ARCHITECTURAL CONSTRAINTS
MAIN PROCESSOR	85.0	Hardware architectural constraints restrict the use of this PLC to SIL 1
POWER SUPPLY	99.8	
ANALOG IN MODULE + 3 CHANNELS	84.2	
DIGITAL OUT LOW MODULE – 2 CHANNELS	90.4	

118   SAFETY EQUIPMENT RELIABILITY HANDBOOK   © 2003 exida.com L.L.C.

EXAMPLE 10-2

PROBLEM: Given the failure rate data from Table 10-2, what is the total failure rate of the PLC portion of the safety instrumented function from Example 10-1?

Solution: Three analog input channels and two digital output channels are required along with all common circuitry. The failure rates for each mode of all components used in the SIF must be added. For each failure mode (using terms from ISA-TR84.00.02, Part 1) the equation is:

$$\lambda_{SIF} = n_{IC} \times \lambda_{IC} + n \times \lambda_{IP} + \lambda_{MP} + m \times \lambda_{OP} + m_{OC} \times \lambda_{OC}$$

where

n_{IC} = number of input channels
n = number of input modules
m = number of output modules and
m_{OC} = number of output channels

The failure rate totals for Example 10-2 SIF are shown in Table 10-3.

Table 10-3. Example 10-2 Failure Rates

Type	Quantity	Per component SD	SU	DD	DU
Analog In Channels	3	25	25	13	38
Analog In Modules	1	850	150	750	250
Main Processor	1	4500	500	3500	1500
Power Supply	1	4513	238	238	13
Digital Out Modules	1	425	75	375	125
Digital Out Channels	2	50	50	25	75
Totals		Totals per SIF			
Analog In Channels		75	75	39	114
Analog In Modules		850	150	750	250
Main Processor		4500	500	3500	1500
Power Supply		4513	238	238	13
Digital Out Modules		425	75	375	125
Digital Out Channels		100	100	50	150
SIF Subtotal		10463	1138	4952	2152
SIF Total		18705			

Exercises

10-1. Can a relay logic system be used to satisfy the requirements for a SIL 3 Safety Instrumented Function? What are some of the issues that need to be addressed in using relays as a logic solver?

10-2. Describe the differences between a 2oo2D and a 1oo2D logic solver. Under what conditions would one consider using a 2oo2D logic solver instead of a 1oo2D logic solver?

10-3. What are the differences between the terms redundancy, voting and architecture?

10-4. A logic solver can be configured to operate either in a 3-2-0 or a 3-2-1-0 mode of operation. Explain the differences.

10-5. A manual emergency shutdown button is required to shutdown a process that uses a Safety PLC as the logic solver. Based on the requirements of IEC 61511, is the emergency shutdown button required to be wired directly to the input of the PLC?

10-6. List the advantages of using programmable electronic systems in safety applications as compared to solid state logic.

10-7. A safety instrumented function uses two discrete inputs and one discrete output. Based on the failure rates of Table 10-2, what is the total dangerous undetected (DU) failure rate for this safety instrumented function?

10-8. What characteristics are considered to be required in a safety PLC?

REFERENCES AND BIBLIOGRAPHY

1. IEC 61508, *Functional Safety of electrical / electronic / programmable electronic safety-related systems*. 2000.

2. *Programmable Electronic Systems in Safety Related Applications, Part 1, An Introductory Guide*. Heath and Safety Executive, 1987.

3. *Programmable Electronic Systems in Safety Related Applications, Part 2, General Technical Guidelines*. Heath and Safety Executive, 1987.

4. DIN V VDE 0801, *Grundsätze für Rechner in Systemen mit Sicherheitsaufgaben*. 1990.

5. DIN V VDE 0801 A1, *Grundsätze für Rechner in Systemen mit Sicherheitsaufgaben, Änderung A1*. 1994.

6. Goble, W.M., "The 1oo2D Safety PLC – How it works," *Proceedings of the Spring ISA Symposium, New Orleans*. ISA, 1997.

7. TR84.00.02-2002, Technical Report, *Safety Instrumented System (SIS) – Safety Integrity Level (SIL) Evaluation Techniques.* ISA, 2002.

8. Goble, W. M., *The Use and Development of Quantitative Reliability and Safety Assessment in New Product Design*, PA: Sellersville, exida, 2000.

11 SIS Final Elements

Final Elements

Many devices have been used as final elements in safety instrumented functions. These include annunciation devices like horns, flashing lights or sirens in functions designed for consequence reduction. Some safety instrumented functions need only simple devices like relays, motor controllers and solenoid valves as the final element. In machine control applications the final element may be a motor controller or a device as complex as a clutch-brake assembly which rapidly removes kinetic energy.

In the process industries the most common final element is a remote actuated valve. This assembly usually consists of a pneumatic/hydraulic control assembly, an actuator and a valve (Figure 11-1). The control assembly may be as simple as a three-way solenoid, a smart partial valve stroke box or a complex electro-pneumatic assembly with solenoids, test switches, pneumatic booster relays and quick exhaust valves.

There are many different types of actuators. They are defined by their power source and range of motion. Actuators typically use electric, hydraulic or pneumatic power sources. The range of motion is typically linear, partial turn or multi-turn.

There are many different types of valves used in safety instrumented functions. Typical valve types include ball, butterfly, offset butterfly, gate, globe and other special designs that are not readily categorized into the standard design groups.

All of these different designs have different performance characteristics, different failure rates and different failure modes.

158   SIS Final Elements

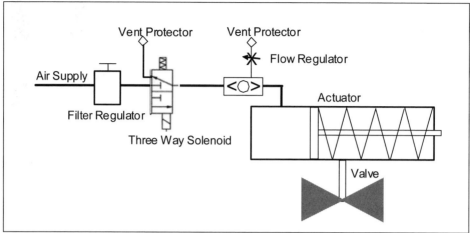

Figure 11-1. Remote Actuated Valve Assembly

The "Well Designed" Remote Actuated Valve

Remote actuated valves must be chosen to match process requirements. Design decisions include materials of construction, valve seat material, valve type, actuator type and controls characteristics. The design must match utility supply levels (air pressure, hydraulic pressure, flow capacity) and tolerances with actuator design to provide the correct torque/thrust to the valve. The torque/thrust must be above breakaway requirements and must be below stress limits of the drive train. With some valves, the ratio of these values may limit the available operating safety factor.

Process material, pressures, temperatures and flow rates will have a major impact of valve type chosen. Matching process needs, including the need for tight shut-off and other safety requirements must also be done. Valve manufacturers often customize the valve design to match specific needs. It is recommended to consult with experts from the valve manufacturer to assure a well-designed final element.

The pneumatic or hydraulic controls must also be carefully designed to not only provide the required torque/thrust levels but also to perform their function in the correct time period. This may require the addition of pneumatic boosters or quick exhaust valves. The final element must close fast enough to provide the safety protection function but not so fast as to induce excessive and possibly dangerous stresses in the piping due to the system dynamics.

Fortunately for the safety instrumented function designer, many valve manufacturers are getting their products and their design processes assessed to IEC 61508 (Ref. 1). These manufacturers supply a safety manual and all failure rate data (typically assuming a well designed final element) needed for SIF probabilistic verification. Some manufacturers are providing complete final elements assemblies carefully designed to meet process requirements.

The concept of a "well designed" remote actuated valve is important. It should be understood that random failure rates and product life can vary by an order of magnitude or more if remote actuated valves are not designed, installed and maintained correctly. That is why the quality and reputation of the manufacturers of these components is so important. A safety manual with complete design, installation and maintenance information can make the difference between success and failure in a safety instrumented function design. There are also specific issues of the integration of the complete assembly. Controls, actuators and valves may be from separate sources with reliability data reflecting the individual component. Depending on the characteristics of the assembly, there may be complementary characteristics of the components that may be exploited when analyzed as an integrated unit.

Actuator Types

The primary grouping of actuators by power source and range of motion permit some convenience in defining the actuator. The conversion of power to motion presents a large range of actuator designs. The different designs have distinctly different functional characteristics to consider within the SIS arena.

Hydraulic and pneumatic piston and diaphragm actuators provide a linear output. These may be integrated with scotch yoke or crank arm mechanism to provide a quarter turn output. These two mechanisms have variations in the output torque over the stroke of the actuator. When integrated with a rack and pinion mechanism they may provide up to a full turn output.

Hydraulic and pneumatic vane actuators provide a direct translation from the vane to a partial turn actuator output without the additional conversion mechanism required for a diaphragm or piston actuator.

Hydraulic and pneumatic motor drives may be configured to provide a range of outputs based on gearing mechanisms. Gearing provides multi-turn and partial turn outputs. The addition of a stem nut system permits translation to linear operation.

Electric actuators have a range of output similarities to the hydraulic and pneumatic motor drives, but typically have a much higher level of complexity due to the additional functionality provided with the units.

A significant issue to address is the impact of "fail safe" versus "non fail safe" actuators. Some designs such as diaphragm and piston actuators with hydraulic and pneumatic power are readily available as spring return valves. The failure or release of control pressure permits the spring to drive the actuator to its safe position.

In some cases a double acting or fail in place actuator may be provided with an accumulator on the pressure to close one side of the piston or diaphragm. This provides a pressure source to return the valve to the safe position in the event of failure of the pressure source that is holding the valve in the operational position. This may be used in areas of limited space or with sizes of actuators that do not readily permit the installation of springs. This situation introduces some additional failure modes in respect to a spring return, but it does provide a protection from the loss of the power source.

Electrically powered actuators have not historically been able to provide fail safe functionality. Today there are some smaller partial turn electric actuators that have a spring return in the event of total loss of power. While these have the ability to close without power, there is a significant addition of complication that impacts the reliability when compared to a simple spring return hydraulic or pneumatic actuator. A response to this has been the electro-hydraulic actuator that permits the safety functionality of a hydraulic piston fail safe actuator. The electrical portion provides power to the hydraulic system and the loss of which may be treated as a trip signal in addition to the basic shutdown signal.

There are advantages in safety instrumented system design to choosing a pre-designed matched set of actuator and valve. A correct combination of actuator and valve will match characteristics of each to provide the best optimization of fail-safe characteristics.

Valve Failure Modes

There is a range of issues to consider with the selection of valves. A valve for SIS service is different from a BPCS valve. The BPCS valve is often selected based on issues such as cycle life and number of repetitive operations. The design of a BPCS valve may have items which actually reduce its reliability in an SIS operation. In some cases a BPCS valve may be protected from certain issues of the SIS valve because the regular operation protects it from failure modes such as media packing or polymerization.

When considered alone as a component, there are no safe failures within a valve. When addressed with a specific type of actuator, some dangerous modes may convert to safe modes. Due to the flow and pressure within the valve, the dynamic and static loads may be used to assist the actuator during tripping. This needs to be carefully addressed for operation and installation since the direction of flow to achieve these characteristics may not be the typical direction of installation for the valve type.

There are a range of failure modes which affect large ranges of valve designs due to shared features. Care must be taken in the analysis of these failure modes. This limits the definition of a particular design as the "best" SIS valve. The dangerous failure modes can be simplified to two basic issues, failure to move to safe position and failure to seal upon reaching safe position.

Failure to Move to Safe Position-Binding

Failure to move to safe position can occur with either a binding or a breakage of some point in the drive train.

Binding may occur between the closure member and the seat or between the stem and the stem bore. The sensitivity of binding between the closure member and the seat depends on the amount of contact the two surfaces maintain. Because of this, ball valves have a higher incidence of this failure compared to a butterfly valve. Binding failures are non-existent in a globe valve.

Stems may also bind in the same manner. Stems with large surface contact area will have significantly more binding failures. The linear stem of a globe valve has a higher binding risk than the quarter turn stem of a ball valve. There are also some risks from stem packing design. There is the risk of over-tightening stem packing during maintenance that can raise operating loads above the capacity of the actuator. With very course media, certain designs may be blocked from full stroke by solids.

Failure to Move to Safe Position – Breakage

There are weaknesses in the valve drive train that may manifest itself when there are increased operating loads in the valve. This is seen more in valves with proportionately longer stems or more point loads in the design. This tends to be seen more in butterfly valve stems compared to the stubbier ball valve stems. Butterfly valves also tend to have more designs with point loads to transfer torque between the stem and the disk.

Failure to Seal Upon Reaching Safe Position

Leakage on completion of stroke is typically caused by damage to the seat (Figure 11-2) or by solids holding the seat and closure member apart. This varies by design and application due to how different designs trap or clear the solids during closure. It is also dependent on the type of solids in the application. Valves with more closure member to seat contact have more opportunity to accumulate damage. Sliding engagement closure members such as ball valves have more possibilities to scar the surfaces during cycling, but the seats are protected during normal flow. Lifting seats such as globe valves have limited scarring exposure during cycling, but the seat and closure members are exposed to the flow during normal operation.

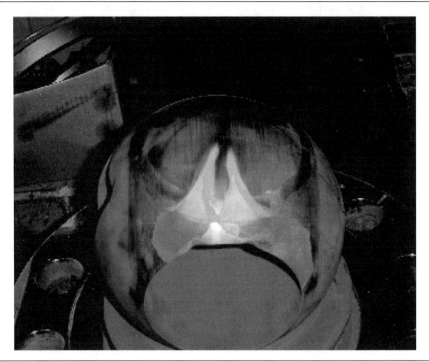

Figure 11-2. Damaged Ball

Valve Types

Trunnion Mount Ball

A ball valve uses a ball rotated ninety degrees to permit or stop process flow. When the valve is opened, a port in the ball provides nearly unobstructed flow to the process materials. When the ball is rotated, it blocks the process flow.

A trunnion mount ball valve has the ball connected to a shaft above and below the ball. The upper shaft transmits the actuator torque to the ball and both shafts together hold the ball in a centralized position against the line pressure. Bearings are typically provided between the trunnion shafts and the valve body. The advantage of this approach is that a lower torque is required to move the ball when compared at similar pressures to a floating ball valve.

The valve has floating seats on the upstream and downstream sides of the ball. In most trunnion designs, the differential pressure on the upstream seat provides the major loading to seal the seat to the ball.

Floating Ball

A floating ball valve has a ball with an upper shaft to rotate the ball for operation. Upstream pressure on the ball pushes the ball into sealing contact with the downstream seat. Since the pressure acting on the ball is the diameter of the seat area and not the more limited seat area of the trunnion valve, floating ball valves normally require more torque to operate than an equivalent pressure trunnion ball valve.

Resilient Butterfly

A resilient butterfly valve offers low cost and low weight. A disk is mounted on a one or two-piece shaft that rotates the disk parallel with the pipe when the valve is open or perpendicular to the pipe when the valve is closed. When the valve is open the disk does provide an obstruction to the flow.

The seats are typically elastomer which is sometimes reinforced with plastic or phenolic backings. They are often available with plastic coating of disks and seats for chemical resistance. Resilient butterfly valves are typically available for lower pressures (up to 250 psig).

High Performance Butterfly

The high performance butterfly valve operates like all butterfly valves with a disk mounted in the middle of the valve that is rotated to open and close the valve. The high performance butterfly is designed with a more complex disk assembly often called "double offset." The first offset is done by locating the stem downstream of the centerline of the disk. This allows an unobstructed 360-degree sealing surface. The second offset moves the stem off the vertical axis of the disk. The combination moves the disk quickly away from the seat as the stem is rotated. This is done to reduce operating torque and extend seat life.

The high performance butterfly valve is available with a variety of seat materials including soft elastomer, polymer or metal (Figure 11-3). This design will typically withstand higher pressures and higher operating temperatures.

Triple Offset Butterfly

The triple offset butterfly valve is also a rotary valve with a cone shaped disk assembly that rotates 90 degrees to open and close the valve. The disk assembly is designed differently with a third off-center offset that provides a characteristic of flow to open or flow to close depending on process material flow direction. A wide range of seat materials is available and the valve is capable of operation at higher pressures and greater temperature ranges depending on materials chosen.

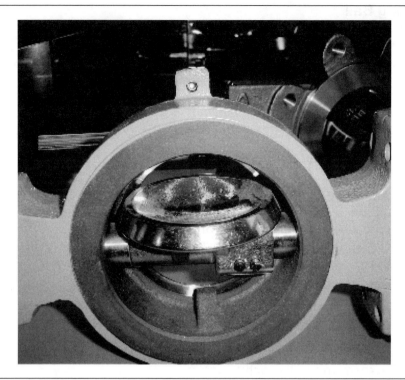
Figure 11-3. Hi-Performance Butterfly Valve

Globe

A globe valve uses a linear stem to lift a plug assembly on and off the seat to permit or stop flow. While many globe valves are fitted with sophisticated trim designs used for various control capabilities, in safety instrumented function applications a simple trim is used for on-off functionality. Various types of seat materials are available with Class IV leakage standard. Tight shut-off to Class VI is available in some manufacturers primarily with soft seat material. Globe valves have operational preferences to flow and pressure direction.

Gate

The simplest functionality of a gate valve is a linear stem that raises and lowers a plate across the ports of the valve. Full conduit valves such as expanding and slab gates provide protection for the seat in the fully open position. Knife and wedge gates permit circulation across the seats in the open position. Valves are available with pressure preferences for fail open or fail closed requirements.

Probabilistic Modeling

Final element assemblies are modeled like any combination of components using the system reliability engineering techniques. Failure mode classification is generally straightforward. The key variables to include in the evaluation of failure rates are:

1. Allowable leakage,
2. Close To Trip versus Open To Trip,
3. Valve and actuator integration,
4. Stress level service category (clean, fouling, severe, etc.), and
5. Diagnostic capability of the system.

Tight Shutoff Requirements

Leakage requirements for valves are specified in IEC 60534-4 (Ref. 2). Different classes of leakage exist with six classes shown in Table 2 of that standard. Class VI is the most stringent with leakage given in terms the number of bubbles per minute allowed during a leakage test. Class IV is a less stringent class with leakage given as 0.1% of rated flow capacity. In many safety instrumented functions, the hazard will be prevented even if the valve leaks a small amount (Class IV, for example). If this level of leakage would not be acceptable, then the valve needs "tight shutoff" characteristics.

Valves that require tight shutoff will have higher failure rates because certain stress events that damage the seat or the ball will be classified a failure. Such events would not be classified as a failure if a small leakage is allowed.

Close To Trip versus Open To Trip

The application of a valve will affect its failure rates and failure modes. One key variable is position of the valve in order for a safety instrumented function to achieve its safe state. The valve must be designed to be "close to trip" or "open to trip." There are generally no tight shut-off requirements when the valve opens to perform the trip.

Service Category

Valves are exposed to highly variable process conditions that affect the failure rates and failure modes of all valves. Two primary categories of service have been used – clean service and severe service.

The term "clean service" is not defined by a standard known to the authors however many manufacturers and reliability analysts have used a definition similar to "clean gas and/or process fluids without (or minimal) particulate or droplets of water."

The term "severe service" has had more varied definitions but a common description is "the valve is exposed to particulates, abrasives and/or corrosive gases or fluids such that process wetted valve components will experience higher stress levels." Example applications include untreated well gas, crude oil and other petrochemical materials.

Failure Rate Comparison

The failure rates of a floating ball valve have been estimated for different application conditions based on a detailed FMEDA of a ball valve. Table 11-1 shows the results.

Table 11-1. Floating Ball Failure Rates as a Function of Application Conditions

Service Conditions	Safe	Dangerous
Clean service, open to trip	0	300
Clean service, close to trip, Class IV leakage allowed	0	400
Clean service, close to trip, Class VI leakage required	0	800
Severe service, open to trip	0	600
Severe service, close to trip, Class IV leakage allowed	0	600
Severe service, close to trip, Class VI leakage required	0	1600

NOTE: All failure rate numbers are in units of failures per billion hours.

It can be seen from looking at the higher failure rate in the last row that it is much harder for the valve to achieve tight shut-off with severe service conditions.

Diagnostics and Annunciation

Like other components in a safety instrumented function, diagnostics can improve the safety and availability performance of a final element. Several diagnostic techniques are given in Table 11-2.

Table 11-2. Remote Actuated Valve Failure Modes and Diagnostic Techniques

Instrument Failure Mode	SIF Failure mode	Diagnostic Technique
Solenoid plunger stuck	Fail-Danger	Pulse test
Solenoid coil burnout	Fail-Safe	Current monitor
Actuator shaft failure	Fail-Danger*	Partial valve stroke test**
Actuator seal failure	Fail-Safe	Pressure monitor
Actuator spring failure	Fail-Danger	Partial valve stroke test**
Actuator structure failure - air	Fail-Safe	Pressure monitor
Actuator structure failure - binding	Fail-Danger*	Partial valve stroke test**
Valve shaft failure	Fail-Danger*	Partial valve stroke test**
Valve external seal failure	No Effect	Unknown
Valve internal seal damage	Fail-Danger	Unknown
Valve ball stuck in position	Fail-Danger	Partial valve stroke test**
* unpredictable - assume worst case		
**assuming an automatic test with pressure wave monitor or position/pressure monitor		

Partial Valve Stroke Testing

There are several different types of "partial valve stroke testing" that have been discussed under a common name. The simplest approach involves a system where the digital output of a PLC is momentarily de-energized in a remote actuated valve that is energized in normal operation. The PLC monitors a switch mounted to detect that the valve assembly has moved as a result of the momentary off pulse.

A more sophisticated approach moves the valve in an analog manner monitoring valve stem position and actuator pressure. Another sophisticated technique pulses the solenoid and measures the pressure response waveform. Certain patterns in the pressure response indicate failures. These methods are capable of doing a better job detecting failures than the simple digital feedback techniques.

Of course the effectiveness depends on many factors including:

- product type used,
- PVST methodology and the analytical capability of the system and users,
- application conditions, and
- tight shut-off requirements.

Care must be taken in implementing PVST, particularly in high-energy conditions. In very high flow conditions the partial stroke test of the SIS valve could incur a response in the BPCS system and the resetting of the PVST could introduce an additional variable to the BPCS. This could impact otherwise unknown instabilities in the BPCS or SIS sensing system leading to a spurious trip.

The effectiveness of any diagnostic including a partial valve stroke test can be determined by a FMEDA. Table 11-3 shows an abbreviated FMEDA for a ball valve.

Table 11-3. Abbreviated FMEDA for a Ball Valve

Valve component	Failure mode and effect	Lambda (FITS)	PST Credit	Lambda (detected
Ball	Damage Pin slot resulting in failure to close	31.50	0	0
Ball	Surface Damage resulting in failure to close	126.00	0.8	100.8
Ball	Surface Damage resulting in Internal leak, major	189.00	0	0
Drive Dowel Pins	Break/damage resulting in failure to close	25.00	0	0
Stem	Break/deformation resulting in failure to close	18.75	0	0
Stem	Damage/Deflection resulting in failure to close	112.50	0.8	90
Trunnion	Damage/Deflection resulting in failure to close	210.00	0.8	168
Seat Seals	Damage resulting in Internal leak, major	20.00	0	0
Stem Bearings	Deformation resulting in failure to close	80.00	0.8	64
Thrust Washers	Deformation resulting in failure to close	20.00	0.8	16
Adapter Bolting	Break resulting in failure to close	1.00	0	0
Adapter Plate Pins	Break/deformation resulting in failure to close	2.00	0	0
Actuator Adapter Plate	Break/deformation resulting in failure to close	1.50	0	0
Seat Ring	Damage resulting in Internal leak, major	180.00	0	0
Seat Ring	Damage resulting in failure to close	60.00	0.8	48
	TOTAL DANGEROUS FAILURE RATE FOR VALVE	1077.25		
	DANGEROUS DETECTED FAILURES			486.8
	TEST EFFICIENCY			45.19%

In Table 11-3 the column "PST credit" identifies the probability of detecting the failure with a partial valve stroke test. The number in this column varies from 0 to 1. A "0" means that no credit is taken. A "1" means that full credit is taken. The dangerous detected failure rate for each component is calculated using the factor in the "PST credit" column. The partial stroke test coverage efficiency is calculated by dividing the dangerous detected failure rate by the total dangerous failure rate. For a ball valve, partial stroke testing provides a diagnostic coverage test effectiveness of 45.19%.

The analysis can be done for valves, actuators or combination actuator/valve devices. Table 11-4 shows an abbreviated FMEDA for a gate valve with piston actuator. This analysis shows that partial valve stroke testing can detect 87.9% of the potentially dangerous failures in this type of product.

Table 11-4. Abbreviated FMEDA of Gate Valve with Piston Actuator

Valve component	Failure mode and effect	Lambda (/Hr)	PST credit	Lambda (detected)
Actuator – Yoke	Deflection causing Failure to close	1.00E-09	1	1.00E-09
Actuator – Cylinder	Damage causing Failure to close	2.50E-09	1	2.50E-09
Actuator – Piston	Damage/Deflection causing Failure to close	1.50E-08	1	1.50E-08
Actuator - Piston Bearing	Damage/Deflection causing Failure to close	4.00E-08	1	4.00E-08
Actuator – Spring	Damage/Deflection causing Failure to close	2.55E-08	1	2.55E-08
Actuator, air - Cylinder Bush	Damage/Deflection causing Failure to close	3.60E-08	1	3.60E-08
Actuator, air - Piston Seals	Damage causing Failure to close	1.00E-08	1	1.00E-08
Actuator, air - Piston Rod Bearing	Deflection causing Failure to close	2.00E-08	1	2.00E-08
Actuator, hyd - Guide Rods	Break causing Failure to close	1.20E-08	0	0.00E+00
Actuator, hyd - Piston Seals	Damage causing Failure to close	1.00E-08	1	1.00E-08
Actuator, hyd - Cylinder Cap	Deflection causing No spring closure	2.50E-09	1	2.50E-09
Actuator, hyd - Spring Plate	Deflection causing No spring closure	2.50E-09	1	2.50E-09
Valve – Body	Deflection causing Failure to close	1.00E-10	1	1.00E-10
Valve – Bonnet	Deflection causing Failure to close	1.00E-10	1	1.00E-10
Valve – Seal Carrier	Damage causing Failure to close	5.00E-10	1	5.00E-10
Valve – Seat Ring	Damage causing Failure to close	4.00E-09	1	4.00E-09
Valve – Seat Ring	Damage causing Internal leak, major	4.00E-09	0	0.00E+00
Valve – Slab Gate	Break Tslot causing Failure to close	7.00E-09	0	0.00E+00
Valve – Slab Gate	Surface Damage causing Failure to close	2.80E-08	1	2.80E-08
Valve – Slab Gate	Surface Damage causing Internal leak, major	1.75E-08	0	0.00E+00
Valve – Stem	Break at THead/DNut causing Failure to close	6.25E-09	0	0.00E+00
Valve – Stem	Damage/Deflection causing Failure to close	3.75E-08	1	3.75E-08
Valve - Bearing Set	Damage causing failure to close	4.00E-08	1	4.00E-08
Valve – Stem Seal Set	Damage causing Failure to close	5.40E-08	1	5.40E-08
Valve, some - Seat Skirt	Deflection causing Failure to close	2.00E-08	1	2.00E-08
	TOTAL DANGEROUS FAILURE RATE	3.96E-07		
	DANGEROUS DETECTED FAILURES			3.E-07
	TEST EFFICIENCY			87.9%

Table 11-5 shows failure rates and coverage factors for the more sophisticated partial valve stroke techniques for different product types determined by FMEDA analysis.

Table 11-5. Partial Valve Stroke Capability

Product Type	Application	Partial Valve Stroke Dangerous Coverage Factor
Solenoid	De-energize to trip	99.0%
Pnuematic Piston Actuator, clean service	De-energize to trip	99.3%
Pneumatic Piston Actuator, severe service	De-energize to trip	99.6%
Pneumatic Rack & Pinion Actuator, clean service	De-energize to trip	81.9%
Pneumatic Rack & Pinion Actuator, severe service	De-energize to trip	88.0%
Scotch Yoke Actuator, clean service	De-energize to trip	92.6%
Scotch Yoke Actuator, severe service	De-energize to trip	94.0%
Gate Valve, clean service	Close to trip	87.9%
Gate Valve, severe service	Close to trip	84.9%
Ball Valve, severe service, full stroke only	Close to trip	45.2%
Ball Valve, severe service, tight shut-off	Close to trip	22.2%
Resilient Butterfly Valve, clean service	Open to trip	63.6%
Resilient Butterfly Valve, clean service	Close to trip	53.8%

Full Valve Stroke Testing

Greater levels of diagnostic coverage can be obtained with full stroke testing. This test could be designed to measure even valve leakage characteristics. One scheme for this testing requires two valves piped in a 2oo2 architecture such that one valve can be completely closed while the other remains open to assure continued process operation. Care must be taken to ensure that the testing accounts for any failure modes in actual operation which the testing is unable to address. If there are particular flow dynamics or differential pressure issues with the valve function, the tested valve may not have this tested as flow is bypassed through the alternate valve.

A number of variations of this design have been done with different levels of instrumentation. The objective of the design is to detect the dangerous failures of the valve assembly. This can be very important especially in SIL3 applications.

Exercises

11-1. What are the most important criteria to use when selecting final elements for safety instrumented function applications?

   a. matching valve capabilities to process requirements
   b. IEC 61508 assessment
   c. purchase cost
   d. manufacturers quality system rating

11-2. Failure rate and failure mode data is available for which final element components:

   a. solenoids
   b. actuators
   c. valves
   d. all components

11-3. When would a designer choose "proven in use" justification in preference to full 61508 assessment?

11-4. What diagnostic coverage would be expected using a high quality partial valve stroke test for a gate valve with piston actuator?

11-5. What diagnostic coverage would be expected using a high quality partial valve stroke test for a pneumatic rack and pinion actuator in severe service conditions?

11-6. Does the need for tight shut off in a valve have an impact on the failure rate?

11-7. Does a valve have the same failure rates and failure modes if used in an open-to-trip configuration versus a close-to-trip configuration?

11-8. What is the primary dangerous failure mode in a ball valve?

REFERENCES AND BIBLIOGRAPHY

1. IEC 61508, *Functional Safety of electrical / electronic / programmable electronic safety-related systems*. 2000.

2. IEC 60534-4, *Industrial-process Control Valves – Part 4: Inspection and Routine Testing*. 1999.

12
Typical SIF Solutions

Introduction

The determination of Safety Integrity Level (SIL) for safety instrumented functions (SIF) is a basic concept of performance based safety standards. The safety integrity or performance of a SIF must increase with higher SIL levels. There are a number of factors that influence the performance of any safety instrumented function. Some of the key factors are:

- Quality of the components used – total failure rate
- Quality of instrument manufacturer to reduce systematic failures
- Safe versus dangerous failure rate of components
- Automatic diagnostic capability of components
- Automatic diagnostics within the SIF
- Proof testability of the components and the SIF
  - Quality of testing, i.e., what % of the component is actually being tested
  - Portion of each component being tested vs. the portion not being tested
- Redundancy of components
- Common cause strength
  - Diversity of redundant components
  - Physical separation of redundant components

- Use of energize to trip vs. de-energize to trip systems.
- Response time of components.
- Time to repair instrument.
- Systematic failures, e.g. failures that relate to the inherent design of the system rather than random hardware failures.
- Safety Lifecycle activities including audits, assessments, and verifications.

In the opinion of committee members on functional safety standards, some of the above factors cannot be practically quantified, e.g., systematic faults like software bugs or procedural errors. Hence functional safety standards provide requirements for protection against systematic faults as well as requirements to do probabilistic calculations to protect against random failures. For the typical SIF solutions being reviewed in this chapter the results of probabilistic SIL verification calculations, including architecture limitations per IEC 61508 (Ref. 1), will be used to demonstrate whether the design satisfies the SIL requirements.

It also needs to be emphasized that SIL is a lifecycle issue and although our present focus is in the hardware solutions required to satisfy the safety functions, all phases of the Safety Lifecycle have to be reviewed for final SIL verification. For all the examples and solutions, a low demand mode of operation is assumed since this is the mode that predominantly applies to the process industries.

The failure rates and other parameters used in the examples are not to be assumed to be typical for all applications. The data used by those who do SIF verification calculations must be evaluated for every application. When stress conditions are above the levels specified in the data reference [for example, (Ref. 2)] then failure rates must be raised.

Although the solutions in this chapter focus on safety integrity evaluation, it must be remembered that the reliability of the system is also important since this impacts on the nuisance or spurious trips of the plant. These trips can have a large impact on safety and cost. Therefore (Mean Time To Fail Spurious) is also calculated.

Since a Safety Instrumented Function is designed to prevent or mitigate a specific hazardous event it is also very important that those involved in SIL verification understand which and how many components have to operate successfully for the function to be effective.

From an architecture and hardware point of view, all items that can fail and prevent proper functioning of the function have to be considered in the SIL verification including the sensors, logic solver and the final elements (actuator). Figure 12-1 provides a more detailed breakdown of the components associated with these main elements.

Typical SIF Solutions   175

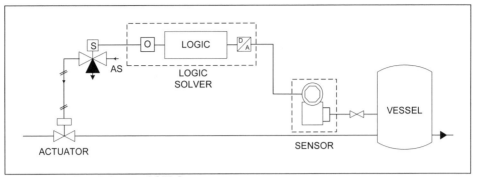

Figure 12-1. Breakdown of SIF Components

In reviewing the above figure it is possible to list all the components which may make up the SIS in the following categories, i.e., Sensor, Logic Solver, Final elements and other devices.

Sensor:

- Process block valve
- Leg lines
- Tracing and insulation of leg lines
- Transmitter
- Transmitter housing and heating/cooling
- Transmitter manifold assembly

Logic solver:

- Backplane
- Power supplies
- Processor
- Input cards
- Output cards
- Communication cards (if involved in safety functionality)

Final Elements:

- Actuator or solenoid valve
- Pneumatic control components
- Digital valve positioner
- Isolation valve

Other devices:

- Safety barriers

- Power supplies
- Interdevice cables
- Terminals
- Air supplies

A SIF is not just a single sensor, logic solver and trip valve. Each component listed above that has an impact on the overall performance and reliability of the SIS must be taken into consideration in the analysis of each safety instrumented function.

Typical SIL 1 Architecture

Figure 12-2 shows an architecture that can be considered to satisfy SIL1 requirements. It consists of a single sensor in clean service, a relay or conventional PLC logic solver, and a single final element in clean service without tight shutoff requirements. Figure 12-3 shows a process application for which this SIL1 architecture may be required. On high pressure in the vessel the ESD valve is required to close. For this design the system is normally fail safe, i.e., during normal operation the three-way solenoid valve is kept continuously energized and air is supplied to the valve to keep it open. Air or power failure will cause the ESD valve to close.

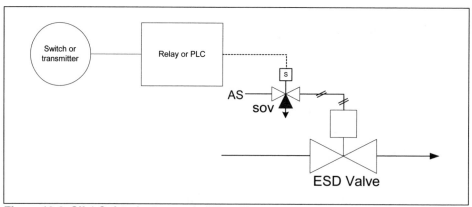

Figure 12-2. SIL1 Safety Instrumented Function

SIL 1 Verification Calculations

Component failure rate data and PFDavg/MTTFS calculations for two cases of the above system are shown in Table 12-1 and Table 12-3. SFF calculations and architecture limit checks are shown in Table 12-2 and Table 12-4. An assumption was made that a very effective proof test was performed at the specified proof test interval with proof test coverage of 95% for the sensor (full calibration), 90% for the logic solver and 80% for final element (full stroke test). The calculation results and all other results

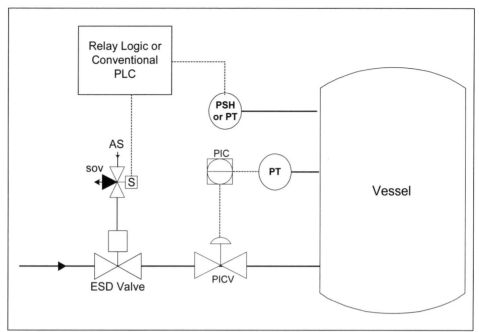

Figure 12-3. Process Application

in this chapter were obtained using exida's SILver™ SIL verification software with major overhaul periods of ten years.

Table 12-1. SIL1 Case 1 - Pressure Switch, Relay Logic, Solenoid Valve, and Trip Valve

ITEM	Failure Rate (per hr) (λ)	% Safe Failure	C^D (%)	PFDavg 1 YEAR TEST INTERVAL (TI)	PFDavg 3 YEAR TEST INTERVAL (TI)	MTTFS (years) @TI=3
Pressure Switch (Clean Service)	6.00E-06	40%	0%	2.24E-02	4.80E-02	49.97
Relay	8.00E-07	70%	0%	1.99E-03	3.69E-03	204.6
Solenoid Valve	6.00E-06	60%	0%			
Shutdown Valve	3.00E-06	55%	0%			
Total for Final Element				4.43E-02	6.59E-02	23.28
TOTAL FOR SYSTEM				6.76E-02	1.14E-01	14.74

The results of the PFDavg calculations in Table 12-1 indicate that for a 1-year test interval the PFDavg is 6.76E-02. This provides a Risk Reduction of 14.8 (SIL1). For a 3-year test interval the PFDavg is 1.14E-01. This provides a Risk Reduction of 8.8. This does not meet SIL 1 requirements. While the one year proof test case does meet the SIL1 limits, there could be some reluctance to implement the function having only a risk reduction of 14.8 for SIL1 applications. Also, if a quantitative SIL selection study was done, it is possible that a Risk Reduction Factor higher than 14.8 was required. In this case the system would not satisfy the requirement. A spurious or nuisance trip is likely to occur every 14.74 years.

Architectural Requirements

To determine the architectural requirements, the SFF number is calculated. This applies to each SIF subsystem, i.e., sensor, logic solver, and final element. To calculate the Safe Failure Fraction for the pressure switch we must first calculate λ^D and λ^{DU}.

$$\lambda^D = (1 - \%S)\lambda$$

$$= (1-0.40) \times 6.00\text{E-}06$$

$$= 3.60\text{E-}06$$

$$\lambda^{DU} = (1 - C^D)\lambda^D$$

$$= (1-0) \times 3.60\text{E-}06$$

$$= 3.60\text{E-}06$$

Therefore,

$$\text{SFF} = \frac{\lambda - \lambda^{DU}}{\lambda}$$

$$= \frac{6.00 \times 10^{-6} - 3.60 \times 10^{-6}}{6.00 \times 10^{-6}}$$

$$= 40\%$$

The Hardware Fault Tolerance (HFT) is 0. This is a Type A device. Therefore, from Figure 7-8, the allowed SIL level can be obtained. In this case it is SIL1. The same process is used for the logic solver and final element. All are Type A devices. The results are summarized in Table 12-2.

Table 12-2. SIL Case 1 - Summary of Architectural Review for Subsystems

Subsystem	Architecture	HFT	SFF	SIL Allowed
Pressure Switch	1oo1	0	40%	SIL 1
Relay	1oo1	0	70%	SIL 2
Solenoid valve	1oo1	0	60%	
Shutdown Valve	1oo1	0	55%	
Final Element	1oo1	0	58.3%	SIL 1

Based on the architectural constraints requirements of IEC 61508 the above system satisfies the requirements for SIL 1 although the PFDavg results are marginal. Note also that both PFDavg calculation and the SFF calculation using the IEC 61508 charts gave the same SIL level. This usually happens when realistic failure rates and proof test parameters are used.

Many consider the use of a conventional PLC for safety applications. In Case 2 a proposed design using a general purpose pressure transmitter, a conventional PLC, a three way solenoid and shutdown valve is being considered. The design is entered into the SIF verification tool and the results are shown in Table 12-3.

Table 12-3. SIL1 Case 2 - Pressure Transmitter, Conventional PLC, Solenoid, and ESD Valve

ITEM	Failure Rate (per hr) (λ)	% Safe Failure	c^D (%)	PFDavg 1 YEAR TEST INTERVAL (TI)	PFDavg 3 YEAR TEST INTERVAL (TI)	MTTFS (years) @TI=3
Pressure Transmitter (clean service)	7.22E-07	65%	40%	3.80E-03	8.25E-03	767.4
General Purpose PLC	2.00E-05	50%	70%	1.64E-02	3.01E-02	10.33
Solenoid Valve	6.00E-06	60%	0%			
Shutdown Valve	3.00E-06	55%	0%			
Total for Valve				4.43E-02	6.59E-02	23.28
TOTAL FOR SYSTEM				**6.36E-02**	**1.02E-01**	**7.09**

The results of the PFDavg calculations for case 2 indicate that for a 1-year test interval the PFDavg is 6.36E-02. This provides a Risk Reduction of 16 (SIL1). For a 3-year test interval the PFDavg is 9.07E-02. This provides a Risk Reduction of 9.8. This does not meet SIL 1 limits. Even the one year proof test interval is marginal and there could obviously be some reluctance to implement the function having a risk reduction of 16 for SIL1 applications. If a quantitative SIL selection study was done, it is possible that a Risk Reduction Factor higher than 16 was required. In this case the system would not satisfy the requirement. The MTTFS for this design is only 7.09 years, and is mainly due to the General Purpose PLC. For most process plants this value is likely to be unacceptable.

Considering the architectural limits, the 3-way solenoid and the shutdown valve are Type A devices. The transmitter and the PLC are both Type B. A summary of SFF calculations and architecture limits are shown in Table 12-4.

Table 12-4. SIL1 Case 2 Architectural Summary

Subsystem	Architecture	HFT	SFF	SIL Allowed
Pressure Transmitter	1oo1	0	79%	SIL 1
General Purpose PLC	1oo1	0	85%	SIL 1
Solenoid Valve	1oo1	0		
S.D. Valve	1oo1	0		
Final Element	1oo1	0	58.3%	SIL 1

Based on the requirements of IEC 61508, the above system satisfies the minimum hardware fault tolerance for SIL 1.

Typical SIL 2 Architecture

Three cases are to be considered to satisfy SIL2 requirements. In all cases a major overhaul time period of ten years was used and "clean service" was assumed. It is also assumed that the safety PLC is programmed to recognize out of range current levels as a diagnostic failure in the associated transmitter and send an alarm. It is also assumed that the safety PLC filters the analog input trip signals so that the analog transition between normal active range and failure current range does not cause a false trip. Under these conditions, all transmitter failures are considered dangerous.

Figure 12-4 consists of a single Safety Transmitter, safety PLC, and single shutdown valve without tight shutoff requirements with a digital valve positioner. The digital positioner would be provided with the capability for Partial Stroke Testing (PST).

In calculating the PFDavg for the trip valve it was assumed that the DVC positioner operated in 0 - 20 mA. mode and operated a pneumatic soft seat ball valve. A diagnostic coverage factor of 70% was assumed for the partial valve stroke testing since tight shut off is not required in the application. To be able to use the 70% diagnostic coverage factor we

Typical SIF Solutions   181

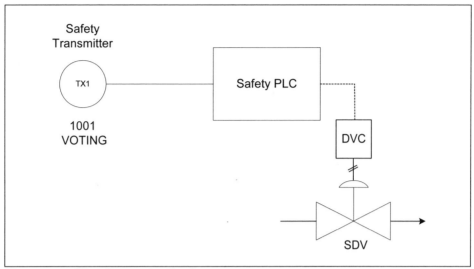

Figure 12-4. SIL 2 Case 1 Application

assumed that the partial stroke test is run at least ten times more often than the expected demand period.

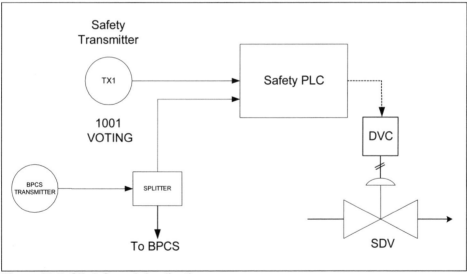

Figure 12-5. SIL 2 Case 2 Application

Figure 12-5 is similar to Figure 12-4 except that instead of just a single Safety Transmitter; the signal from the safety transmitter is compared with the BPCS transmitter in the safety PLC. This example is intended to demonstrate the advantages and benefits of taking credit for improved diagnostics.

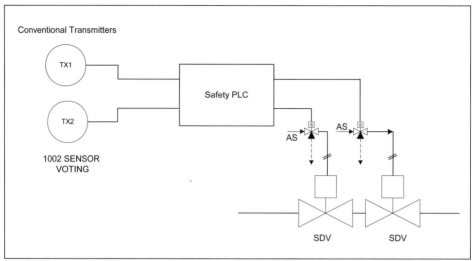

Figure 12-6. SIL 2 Case 3 Application

Figure 12-6 consists of two conventional Transmitters (1oo2 voting), a safety PLC, and two separate shutdown valves (1oo2) with individual solenoids.

Table 12-5. PFDavg Calculations Results for SIL2 Case 1

ITEM	Failure Rate (per hr) (λ)	% Safe Failure	C^D (%)	PFDavg 1 YEAR TEST INTERVAL (TI)	PFDavg 3 YEAR TEST INTERVAL (TI)	MTTFS (years) @TI=3
Safety Transmitter (clean service)	9.12E-07	0%	92%	4.69E-04	1.02E-03	1843
Safety PLC	1.56E-05	74%	94%	4.68E-05	9.56E-05	202
DVC	7.28E-07	95%	0%			
Shutdown Valve	3.00E-06	55%	70%			
Total for valve				5.10E-03	7.70E-03	49.19
TOTAL FOR SYSTEM				**5.61E-03**	**8.80E-03**	**38.73**

The results of the PFDavg calculations for SIL 2 Case 1 indicate that for a one-year test interval the PFDavg is 5.61E-3. This provides a Risk Reduction of 178 (SIL2). Figure 12-7 shows the PFDavg contribution of each subsystem. It can be seen that the final element subsystem is the primary limit. For a three-year test interval the PFDavg is 8.80E-03. This provides a Risk Reduction of 114 (SIL2).

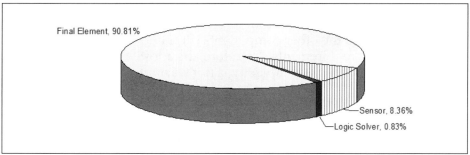

Figure 12-7. PFDavg Contribution of Subsystems – SIL 2 Case 1

Table 12-6. SIL 2 Case 1 Architectural Summary

Subsystem	Architecture	HFT	SFF	SIL Allowed
Safety Transmitter	1oo1	0, Type B	94.6	SIL 2
Safety PLC	1oo2D	1, Type B	99.60	SIL 3
DVC	1oo1	0, Type A		
Shutdown Valve	1oo1	0, Type A		
Final Element	1oo1	0, Type A	88%	SIL 2

Table 12-6 shows that, based on the requirements of IEC 61508, the SIL 2 Case 1 system satisfies the minimum hardware tolerance for SIL 2.

Figure 12-5 illustrates a second example using two transmitters. However, only one transmitter is part of the safety instrumented function equipment set. For this application signal comparison to achieve high diagnostics is done in the Safety PLC. It is also recommended that the BPCS transmitter be hardwired to the Safety PLC. This hardwiring could be via a signal splitter as shown.

The signal comparison allows the designer to take credit for the comparison of the two input signals. A deviation alarm will alert the operator to take corrective action or a safe shutdown can be initiated. External comparison is highly effective for analog signals since one can monitor differences in the dynamic signals and see if something is wrong with one of the signals.

The analog signal comparison can be performed between two or more transmitters that are part of the SIF or between a SIF transmitter and a DCS transmitter (measuring the same process variable). In order to take credit for the comparison it needs to be performed in the SIS logic solver. The software in the Safety Instrumented System is safety critical rated. This is not the case for the DCS software.

IEC 61508 allows a claim of up to 99% for comparison diagnostics. For the above application where it was assumed that the comparison diagnostic will only alarm and not initiate a shutdown, a diagnostic credit of 95% will change the PFDavg for the sensor portion from 1.02E-3 to approximately 5.7E-5 for a three year proof test interval. The function will still be SIL 2 rated due to the other components (PLC and Valve) and architectural constraint requirements. Therefore, the additional equipment is not useful and the additional complexity is not warranted.

In Figure 12-6 the third proposed design is shown. This case consists of two conventional transmitters (1oo2 voting) with comparison diagnostics, a safety PLC, and two separate shutdown valves (1oo2) with individual solenoids. The calculation results are shown in Table 12-7 using a common cause beta factor of 5% for the sensors and 10% for the final elements.

Table 12-7. PFDavg Calculation Results for SIL2 Case 3

ITEM	Failure Rate (per hr) (λ)	% Safe Failure	C^D (%)	PFDavg 1 YEAR TEST INTERVAL (TI)	PFDavg 3 YEAR TEST INTERVAL (TI)	MTTFS (years) @TI=3
Pressure Transmitters (1oo2, clean service)	7.22E-07	65%	40%	1.01E-5	2.61E-5	390.44
Safety PLC	1.56E-05	74%	94%	4.82E-05	9.85E-05	197.35
Solenoid Valve	6.00E-06	60%	0%			
Shutdown Valve	3.00E-06	55%	0%			
Total for Valve				6.95E-03	1.20E-2	11.33
TOTAL FOR SYSTEM				**7.01E-03**	**1.21E-2**	**11.1**

The results of the PFDavg calculations for SIL 2 Case 3 indicate that for a one-year test interval the PFDavg is 7.01E-03. This provides a Risk Reduction of 143 (SIL2). The final elements contribute a significant majority to the PFDavg.

For a three-year test interval the PFDavg is 1.20E-03. This provides a Risk Reduction of 82 which no longer meets the requirements of SIL2. A spurious or nuisance trip is likely to occur every 11.1 years.

The architectural constraint limits are listed in Table 12-8. None of the SIL levels from this table are the limiting factor as is often the case. Based on the requirements of IEC 61508 the above system satisfies the minimum hardware tolerance for SIL 2.

Table 12-8. SIL 2 Case 3 Architectural Summary

Subsystem	Architecture	HFT	SFF	SIL Allowed
Pressure Transmitter (1oo2, clean service)	1oo2	1, Type B	79%	SIL 2
Safety PLC	2oo3	1, Type B	99.6%	SIL 3
1oo2 Solenoid Valve	1oo2	1, Type A		
1oo2 Shutdown Valve	1oo2	1, Type A		
Final Element	1oo2	1, Type A	58%	SIL 2

Typical SIL 3 Architecture

Figure 12-8 and Figure 12-9 show architectures normally considered to satisfy SIL3 requirements. Figure 12-8 consists of two safety transmitters, a safety PLC, and two shutdown valves with a digital valve positioner. Both digital positioners would be provided with capabilities for Partial Stroke Testing (PST).

As with the SIL2 architecture when using PST, in calculating the PFDavg for the shutdown it was assumed that the DVC operated in 0 - 20 mA mode operating a pneumatic valve. For the shutdown valve a diagnostic coverage factor of 70% was used. To be able to use the 70% diagnostic coverage factor we assumed that the partial stroke test is run at least ten times more often than the expected demand period.

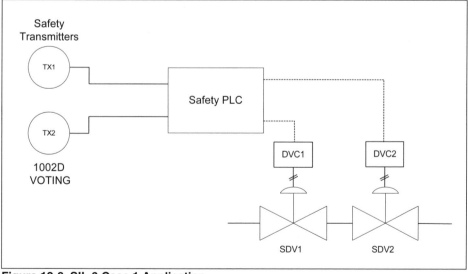

Figure 12-8. SIL 3 Case 1 Application

Table 12-9. PFDavg Calculation Results for SIL3 Case 1

ITEM	Failure Rate (per hr) (λ)	% Safe Failure	C^D (%)	PFDavg 1 YEAR TEST INTERVAL (TI)	PFDavg 3 YEAR TEST INTERVAL (TI)	MTTFS (years) @TI = 3
Safety Transmitters (1oo2D, clean service)	9.12E-07	0%	92%	4.64E-07	1.01E-06	6279
Safety PLC	1.56E-05	74%	94%	4.82E-05	9.57E-06	197.35
DVC	7.28E-07	95%	0%			
Shutdown Valve	3.00E-06	60%	60%			
Total for Valve				5.44E-04	8.47E-04	25.87
TOTAL FOR SYSTEM				5.93E-04	9.47E-04	22.79

The results of the PFDavg calculations for SIL 3 Case 1 indicate that for a one-year test interval the PFDavg is 5.93E-04. This provides a Risk Reduction of 1684 (SIL3). For a three-year test interval the PFDavg is 9.47E-04. This provides a Risk Reduction of 1,056 (SIL3). Both values are within the acceptable bands for SIL3 although the three-year test interval is a little marginal. A spurious or nuisance trip is likely to occur every 22 years. The architecture constraints also allow SIL 3.

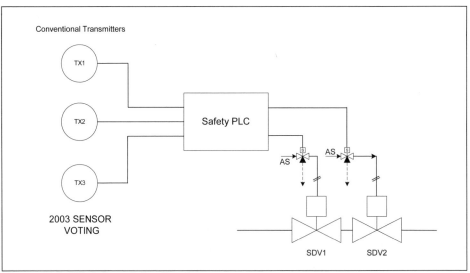

Figure 12-9. SIL 3 Case 2 Application

In the SIL 3 Case 2 design, three conventional transmitters are connected to the safety PLC and the two valve assemblies use three way solenoids instead of digital valve positioners. No partial valve stroke testing is done. The results of the PFDavg calculations are listed in Table 12-10.

Table 12-10. PFDavg Calculation Results for SIL3 Case 2

ITEM	Failure Rate (per hr) (λ)	% Safe Failure	C^D (%)	PFDavg 1 YEAR TEST INTERVAL (TI)	PFDavg 3 YEAR TEST INTERVAL (TI)	MTTFS (years)
Pressure Transmitters	7.22E-07	65%	40%	1.02E-05		
Safety PLC	7.78E-06	60%	99%	4.88E-05		
Solenoid Valve	6.00E-06	60%	0%			
S.D. Valve	3.00E-06	55%	0%			
Total for Valve				6.95E-03		
TOTAL FOR SYSTEM				**7.01E-03**		

The results of the PFDavg calculations for SIL 3 Case 2 indicate that for a one-year test interval the PFDavg is 7.01E-03. This provides a Risk Reduction of 143 (SIL2). The relative subsystem contribution is shown in Figure 12-10. This clearly shows that the final element design must be improved to achieve a SIL 3 rating for the safety instrumented function. The three-year proof test interval was not calculated.

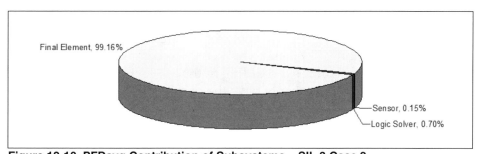

Figure 12-10. PFDavg Contribution of Subsystems – SIL 3 Case 2

Since the PFDavg requirements have not been met, it is likely that the architecture or technology used may have to be changed. It is therefore of no value at this point to investigate whether the hardware fault tolerance requirements are met.

Some Common Hardware Issues Relating to the Various Solutions for SIL1, SIL2, and SIL3 Systems

Some of the issues relating to the various architectures are:

1. Use of Transmitters instead of Switches – For use in safety systems the use of transmitters instead of switches is highly recommended for the following reasons:
   - Better reliability
   - Diagnostics
   - Cost
   - Ease of maintenance and testing

2. Use of Safety vs. conventional transmitters – Safety transmitters are inherently designed for safety application and as such the percentage of safe versus dangerous failures is high. Also, even for the dangerous failures the diagnostic coverage factor, i.e., the ability of the transmitter to detect that a dangerous failure has occurred is also very high.

3. Use of multiple SIS: Multiple Safety Instrumented Systems are sometimes used to provide a higher SIL capability, e.g., the use of two SIL1 systems to possibly satisfy the needs for a SIL2 system. Generally this will not work. When using multiple SIS a careful analysis is required to ensure that maintenance, common cause issues are properly analyzed. There is also the issue of "SIL capability." When a PLC is certified, it must follow a rigorous quality process according to the SIL level (see Chapter 7). A SIL 1 PLC will not likely have the quality needed for SIL 2. A SIL 2 PLC will not likely have the quality needed for SIL 3.

4. For the redundant architectures outlined for the various SIL levels, common cause failures have to be reviewed since these failures have a major impact on the overall performance and integrity of the systems.

REFERENCES AND BIBLIOGRAPHY

1. IEC 61508, *Functional Safety of electrical / electronic / programmable electronic safety-related systems*. 2000.

2. *Safety Equipment Reliability Handbook*. exida, 2003. (available from ISA)

13
Oil and Gas Production Facilities

Introduction

Oil and Gas production facilities consist of many diverse and complex processes that have special safety challenges that are similar to and also significantly different from those in Chemical and Petrochemical plants. These processes must run with the highest possible reliability under differing and difficult operating terrain that changes from permafrost to desert conditions to deep seas. Under these harsh conditions, continuous and safe operation is expected 24 hours per day, 7 days per week with negligible impact to the environment. The remote location of the facilities also creates special maintenance and operating challenges.

The processes mainly consist of:

- Offshore platforms that are structures erected in seas primarily for the extraction of petroleum products.

- Onshore oil and gas fields with wells and associated controls.

- Tank farms and underground storage facilities that may consist of salt caverns or porous layers underground.

- Oil and Gas treatment facilities. Once pumped from a well, the oil and gas has to be separated from other contaminants such as water, sulfur, additives and other hydrocarbons that harm equipment and the environment or reduce the quality and usability of the product. Installations for carrying out these treatments can vary considerably.

- Pumping systems for the transportation of liquids.

- Compression facilities for transportation of gases.

- Injection facilities for injecting gas and liquids into wells and storage facilities. Since production pressure from wells tends to decline over time, hydrocarbons are re-injected to maintain or increase the pressure.

- Pipelines for the transportation of various hydrocarbon products using pipeline systems, compression facilities, pumping systems, valve stations, storage and distribution facilities.

Oil and gas wells, whether onshore or offshore, operate at extremely high pressures, flows, and temperatures, and the hazards and relating consequences due to operational or equipment problems can be catastrophic, leading to multiple loss of life, substantial revenue losses, and major environmental incidents.

Companies involved with oil and gas exploration and production from both onshore and offshore facilities, therefore, face the same cost, safety, production, and environmental issues as other industries worldwide.

New challenges are being created for those involved in the design of the facilities. The areas in which the facilities are being installed are becoming more remote. With increasing social and economic pressures, the legislation in various countries is becoming more stringent. These issues are addressed to a great extent by ensuring that the safety instrumented systems are specified, designed, installed, operated, and maintained as per appropriate and well recognized safety standards.

The design of safety systems for wells has been and is still very prescriptive in nature based on current recommended practices like API RP 14C (Ref. 1). Many companies today are also reviewing the design of the safety system for the wells for compliance with the performance-based requirements of IEC 61508 (Ref. 2) and ANSI/ISA-84.00.01-2004 (IEC 61511 Mod) (Ref. 3) and comparing the differences. Most are discovering that the new standards allow higher levels of safety at a lower cost, thus the enthusiasm for the new standards.

In this chapter a Safety Instrumented Function (SIF) for gas wells encountered in gas fields will be reviewed. For the SIF identified, a detailed review of the function will be done and a calculation of possible PFD_{avg} values achieved will be done so that the SIL verification for the SIF can be accomplished in part.

NOTE: The data and facilities associated with the example described in this chapter have been tailored to enhance the understanding of the SIL verification techniques and procedures. The facilities do not necessarily comply with national or international norms and should not be used as the basis for any design.

Overall System Description

Figure 13-1 is a simplified flow diagram showing the interconnection of gas wells to a treatment plant. In this example, five gas producing wells are located in a gas field. The wells are approximately 2 to 10 km away from the treatment plant. Well pressure varies from approximately 3000 psi (200 barr) to 6000 psi (400 barr).

An adjustable choke is located at each well for adjusting the well outlet pressure. The chokes are controlled at a discharge pressure of 2000 psi (136 barr). A wellhead control panel is located at each well. All critical controls for each well are included in the local panel. All well operation is carried out locally, but key parameters are monitored by operators from the treatment plant main control center. The data from each well to the main control center is sent via a radio communication system.

All the wells are connected to a main production header, a test header, and to a main flare header. The relief valves are connected to the main flare header. For simplification, the flare header is not shown in Figure 13-1. The test header enables personnel to isolate and monitor the production from individual wells.

The main and the test headers are connected to separators. The purpose of the separators is to separate condensate and other liquid contaminants from the gas. The gas and liquids from the separators are further treated before being distributed offsite to customers.

Some critical parameters associated with the system operation are:

- High pressure possibly due to closing of a separator inlet shutdown valve (SDV) or other block valves, improper setting or failure of the choke, or blocking in of a section of line.
- Low pressure due to line rupture or a flange leak
- High temperature of product from well, or high ambient temperature
- Low ambient temperature causing freeze up and possible equipment failure.
- High flow due to excessive demand or leakage/rupture of line.

Individual Well Controls and Shutdowns

Figure 13-2 shows the main components and their interconnections for the basic process control systems and the safety instrumented systems associated with a well. Details of the systems for a well are:

- There are 5 trip valves at each well.

192   Oil and Gas Production Facilities

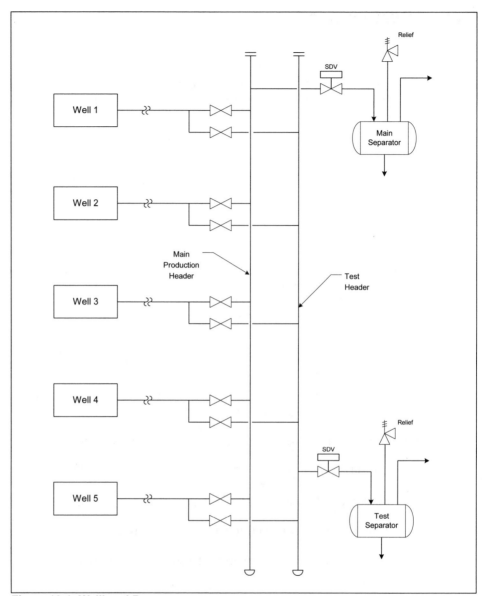

Figure 13-1. Wellhead Process

- A down-hole safety valve (DSV) is located underground. This is a self-contained valve that operates on high flow and does not require external signals or power for its operation. The valve is normally open, and closes automatically when its differential pressure reaches a predetermined value. Based on the flowing capabilities of the well, each valve is adjusted to accommodate the fluid, pressure, temperature and flow rate of the well.

- A lower master trip valve (LMV) is located above ground where the line exits the well. This valve is "fail close" and is hydraulically operated via a solenoid valve.

- An upper master valve (UMV) is located above the LMV and is also "fail close" and hydraulically operated via a solenoid valve.

- A wing valve (WV) is located on the exit line from the well. This valve is also "fail close" and hydraulically operated via a solenoid valve.

- A flow-line valve (FLV) is located on the exit line from the well after the choke. This valve is also "fail close" and hydraulically operated via a solenoid valve.

The lower master, upper master, wing, and flow-line valves are hydraulically operated. Loss of hydraulic pressure will cause the valves to close.

In a proposed design each well has two independent shutdown systems, i.e., ESD1 and ESD2.

- ESD1 is located at the well and consists of pressure switches, relay logic, and the flow-line valve. This trip system closes the flow line valve automatically if the line pressure drops below or goes above set limits.

- ESD2 is independent of ESD1 and uses a safety PLC as the logic solver. This PLC is located at the well. The input trip signals connected to the PLC are confirmed fire, low line pressure, high line pressure, local manual shutdown, remote manual shutdown, and closure of the main separator inlet shutdown valve. Two pressure transmitters in a 1oo2 voting configuration are used for the low and high pressure sensing. Open and Close position switches are located on the main separator inlet shutdown valve. The switches send a signal to the safety PLC at all wells via a radio system to trip all the wells if any of the two switches sense a "not open" or "close" status.

Associated with each well are numerous safety instrumented functions. Some of the common functions are:

- High line pressure
- Low line pressure
- High flow
- Low flow
- Gas detection

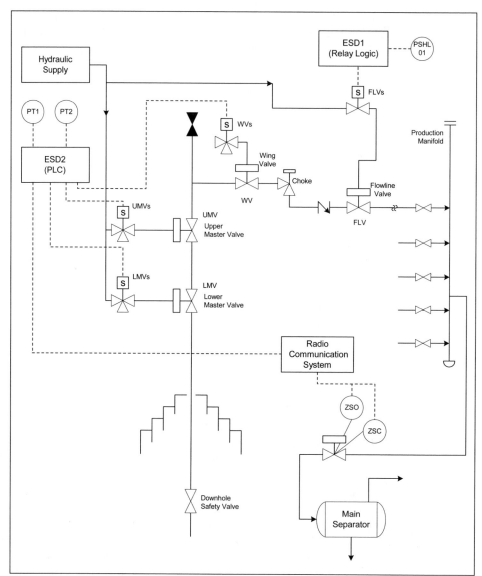

Figure 13-2. Control Panel

- Fire detection
- Local manual ESD
- Remote manual ESD

Prior to carrying out SIL verification calculations for Safety Instrumented Functions (SIF) it is essential and very important that each function and its associated input and output signals be well defined. The identification of the SIF is part of the analysis phase and the detail requirements for each function are documented in the Safety Requirements Specifications (SRS).

The SRS must include the hazardous event and the I/O associated with the each individual function.

High Line Pressure Safety Instrumented Function (SIF)

The focus in this chapter will be on one case of high line pressure in the lines from the wells to the manifold header. High pressure in the lines to and including the manifolds can be due to many causes. Examples are:

1. Closure of the main shutdown valve at the inlet to the main separator. The closure of this valve will cause the lines from all the wells, the manifolds, and the line from the manifold to the main shutdown valve to overpressure. The most credible consequence associated with this scenario (to be based on a detailed PHA review) may be a rupture of a flange in the main manifold assembly leading to a fire and equipment damage.

2. Closure of a main inlet block valve to the manifold header. The closure of this valve will cause the lines from the associated well to the inlet of the valve to overpressure. The manifold and its outlet lines to the main separator S/D valve will not overpressure in this case. The most credible consequence associated with this scenario may be a rupture of the main line from the well to the manifold inlet valve leading to a fire, equipment damage, and loss of life because the line runs parallel to a main highway.

These two causes will result in the definition of two different safety instrumented functions because the consequences and likelihood associated with the two scenarios are different. These differences are summarized in the SIF list for well #1 (Table 13-1).

Figure 13-3 shows a block diagram of the proposed design for the safety instrumented function of the high line pressure. It consists of two separate and independent shutdown functions in two separate and independent systems (ESD1 and ESD2). The components associated with the two systems for this function are:

- **Sensors** – For ESD1, the sensor is a pressure switch. For ESD2, two pressure transmitters and two position switches mounted on the separator inlet valve are sensors. The pressure transmitters and limit switches are configured for 1oo2 voting. The position switches are included as part of this safety function since there is a direct correlation between the closing of the separator inlet valve and high line pressure.

- **Logic** – For ESD1, relay logic is used. For ESD2, a safety PLC is installed.

Table 13-1. SIF Descriptions

SIF	SIF INITATORS	INTERMEDIATE AND HAZARDOUS EVENT	INPUT SIGNALS	SIL	SIF OUTPUTS	AUXILIARY OUTPUTS
13.1 Case 1	High pressure at outlet of well	High pressure mainly due to closure of the main S/D valve at the inlet to the main separator. The closure of this valve will result in all lines from the wells and the manifold to over pressuring. This will likely result in a flange rupture at the main manifold assembly leading to a fire and equipment damage	PSHL-01 -------------- PT-1, PT-2, ZS-1, ZS-2**	2	FLV-01 Closes ---------------- LMV-1, UMV-1, WV-1	None
13.1 Case 2	High pressure at outlet of well	High pressure mainly due to the closure of the main inlet block valve to the manifold header. The closure of this valve will cause the line from the individual well to overpressure resulting a rupture of the main line from the well to the manifold inlet valve leading to a fire, equipment damage, and loss of life	PSHL-01 -------------- PT-1, PT-2	1	FLV-01 Closes ---------------- LMV-1, UMV-1, WV-1	None

**Note: Although we have two separate Safety Systems (ESD1 and ESD2), components of both ESD systems will be used in analyzing both SIF. In Case 1, the position switches on the main S/D valve can be used as an input to the SIF since we know that closing of the valve will lead to high pressure. For Case 2, we cannot take credit for the position switches since the closing of the inlet valve is not the cause to the high pressure for this scenario. We will be analyzing Case 1 in our analysis.

- **Valves** – For ESD1, the flow-line valve is the final element. For ESD2, the lower master, upper master, and wing valves are configured for 1oo3 voting.

SIF PFDavg Calculation

Simplified equations and fault tree analysis will be used to determine the PFDavg of the SIF shown in Figure 13-3. One advantage in using the fault tree approach for the PFDavg calculation for this SIF is that most engineering personnel are familiar with the development and analysis of fault trees. This enables them to better understand the operation of complex safety instrumented functions by reviewing its associated fault tree.

Table 13-2 shows the data used for the various components in calculating the PFDavg for the Case 1, high line pressure safety instrumented function. In addition to this data the "mission time" of the processing unit must be known. For process equipment this is the operating time interval between major turnarounds where the SIF equipment is completely tested, re-built and restored to like new condition. The example will use ten years.

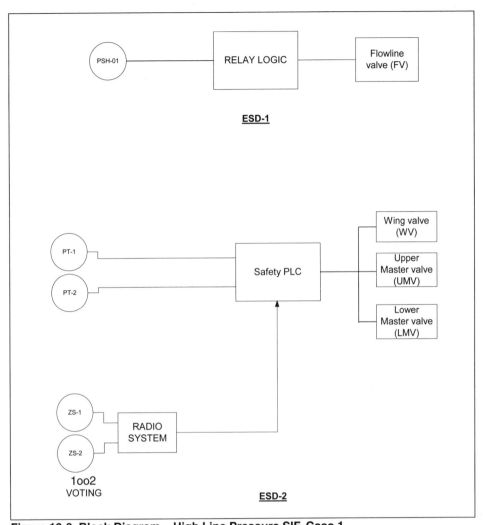

Figure 13-3. Block Diagram – High Line Pressure SIF, Case 1

The manual proof test interval must also be known. The example will use a one-year manual proof test interval.

Generic failure rates were chosen from the *Safety Equipment Reliability Handbook* (Ref. 4) and referenced. For the PLC, a "Generic SIL2" safety PLC was chosen with a single channel architecture.

For each component the total failure rate, Lambda, is given as well as the percentage of the percentage of the total failure rate that is safe. The diagnostic coverage factor for any automatic diagnostics within the SIF is estimated or obtained from references. A restore time is estimated for each component. A "manual proof test effectiveness" (diagnostic coverage of the manual proof test) is also estimated. Using these numbers, three dangerous failure rates are calculated.

Table 13-2. Component Failure Rate Data - Well Trip System

ESD1	Lambda	%Safe	Automatic Test Diagnostic Coverage CD	Repair Time (RT)	Proof Test Coverage	Lambda DD		Lambda DU E		Lambda DU U		Reference
Pressure switch	6.00E-06	40%	0%	72	95%	$\pi_1 =$	0.00E+00	$\pi_{17} =$	3.42E-06	$\pi_{33} =$	1.80E-07	Fig. 8-2
Impulse line	2.50E-07	0%	0%	72	95%	$\pi_2 =$	0.00E+00	$\pi_{18} =$	2.38E-07	$\pi_{34} =$	1.25E-08	EXI-SILver LOW
Relay	1.50E-06	60%	0%	72	90%	$\pi_3 =$	0.00E+00	$\pi_{19} =$	5.40E-07	$\pi_{35} =$	6.00E-08	[EXI03] pg.99
Solenoid valve FLV	6.00E-06	60%	0%	72	80%	$\pi_4 =$	0.00E+00	$\pi_{20} =$	1.92E-06	$\pi_{36} =$	4.80E-07	[EXI03] pg.155
Ball Valve FLV	3.00E-06	25%	0%	168	70%	$\pi_5 =$	0.00E+00	$\pi_{21} =$	1.58E-06	$\pi_{37} =$	6.75E-07	[EXI03] pg.172
ESD2						π		π		π		
Transmitters PT1/PT2	1.50E-06	10%	56%	72	95%	$\pi_6 =$	7.50E-07	$\pi_{22} =$	5.70E-07	$\pi_{38} =$	3.00E-08	Fig. 8-3
Common Cause Beta Factor for Pressure Transmitters π_{PT}	0.02											
Safety PLC Analog Input Channel (IC)	1.02E-07	50%	94%	24	100%	$\pi_7 =$	4.80E-08	$\pi_{23} =$	3.00E-09	$\pi_{39} =$	0.00E+00	[EXI03] pg.124
Safety PLC Analog Input Processing (IP)	2.00E-06	50%	90%	24	100%	$\pi_8 =$	9.00E-07	$\pi_{24} =$	1.00E-07	$\pi_{40} =$	0.00E+00	[EXI03] pg.124
Safety PLC Main Processing (MP)	1.25E-05	74%	95%	24	100%	$\pi_9 =$	3.10E-06	$\pi_{25} =$	1.50E-07	$\pi_{41} =$	0.00E+00	[EXI03] pg.124
Safety PLC Digital Output Processing (OP)	1.00E-06	80%	95%	24	100%	$\pi_{10} =$	1.90E-07	$\pi_{26} =$	1.00E-08	$\pi_{42} =$	0.00E+00	[EXI03] pg.124
Safety PLC Digital Output Channel (OC)	1.20E-07	75%	99%	24	100%	$\pi_{11} =$	2.97E-08	$\pi_{27} =$	3.00E-10	$\pi_{43} =$	0.00E+00	[EXI03] pg.124
Solenoid valves UMV/LMV/WV	6.00E-06	60%	0%	168	70%	$\pi_{12} =$	0.00E+00	$\pi_{28} =$	1.68E-06	$\pi_{44} =$	7.20E-07	[EXI03] pg.155
Ball Valves UMV/LMV/WV	2.45E-06	67%	0%	168	70%	$\pi_{13} =$	0.00E+00	$\pi_{29} =$	5.60E-07	$\pi_{45} =$	2.40E-07	EXI-SILver
Common cause Beta factor for valves	0.1				95%	$\pi_{14} =$	0.00E+00	$\pi_{30} =$	9.50E-02	$\pi_{46} =$	5.00E-03	[IEC00]
Position switch on ZS1/2 on SDV	6.00E-06	40%	0%	72	95%	$\pi_{15} =$	0.00E+00	$\pi_{31} =$	3.42E-06	$\pi_{47} =$	1.80E-07	Fig. 8-2
Common cause Beta factor for position switch	0.1											
Radio system	5.00E-07	50%	30%	72	90%	$\pi_{16} =$	7.50E-08	$\pi_{32} =$	1.58E-07	$\pi_{48} =$	1.75E-08	Author Estimate

Lambda DD represents the failure rate detected by automatic diagnostics within the SIF. Assuming that the system is not designed to automatically shutdown the process when a diagnostic alarm occurs, these failures will cause the SIF to be failed dangerously only during the restore time. Lambda DU E represents the failure rate undetected by automatic diagnostics but detected by the manual proof test. Lambda DU U represents the failure rate not detected by automatic diagnostics and not detected by the manual proof test. If these failures occur, a SIF may fail dangerously for the entire mission time.

Figure 13-4 is the fault tree for the safety instrumented function. It shows that the SIF will fail dangerously only when both ESD1 and ESD2 fail dangerously.

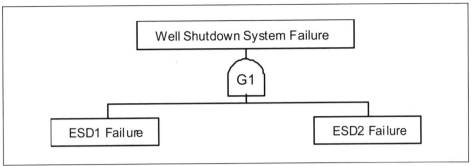

Figure 13-4. Fault Tree for Well Shutdown System

A fault tree for ESD1 is shown in Figure 13-5. It shows that ESD1 will fail

- When the pressure switch fails dangerously but detected by the effectiveness of the proof test (PSH01 DU E),

- OR when the pressure switch fails dangerously and not detected by the proof test (PSH01 DU U),

- OR when the pressure switch impulse line fails dangerously but detected by the effectiveness of the proof test (IMP DU E),

- OR when the pressure switch impulse line fails dangerously and not detected by the proof test (IMP DU U),

- OR when the relay fails dangerously but detected by the effectiveness of the proof test (RELAY DU E),

- OR when the relay fails dangerously and not detected by the proof test (RELAY DU U),

- OR when the FLV solenoid fails dangerously but detected by the effectiveness of the proof test (FLVs DU E),

- OR when the FLV solenoid fails dangerously and not detected by the proof test (FLVs DU U),

- OR when the FLV ball valve fails dangerously but detected by the effectiveness of the proof test (FLV DU E),

- OR when the FLV ball valve fails dangerously and not detected by the proof test (FLV DU U).

ESD1 PFD Calculation

The probability of failure on demand (PFD) for ESD1 can be simply calculated using the unreliability function (Chapter 4) given that the failure rates for failures detected by the proof test are separated from those not detected by the proof tests.

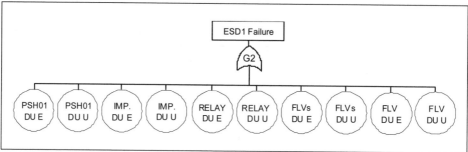

Figure 13-5. Fault Tree for ESD1

Using Equation 4-5 the PFD is:

$$\text{PFD}_{G2} = 1 - \exp(-(\lambda_{17} + \lambda_{18} + \lambda_{19} + \lambda_{20} + \lambda_{21}) \times 8760) = 0.0652$$

for failures detected by the proof test

$$\text{OR } 1 - \exp(-(\lambda_{33} + \lambda_{34} + \lambda_{35} + \lambda_{36} + \lambda_{37}) \times 87600) = 0.1159$$

for failures not detected by the proof test.

Therefore, PFD = 1 − (1 − 0.0652) × (1 − 0.1159) = 0.1736

This can be calculated as a function of operating time interval. A chart of this is shown in Figure 13-6. The effect of the proof testing is shown when the PFD drops every year. The PFD does not drop to zero because the proof test is not perfect. The effect of the imperfect proof testing is shown as the PFD accumulates to its high value. This result is surprising to those who have using simplified equations that do not account for proof test coverage. Although we have chosen relatively high values for proof test coverage, 70% to 95%, the impact is drastic.

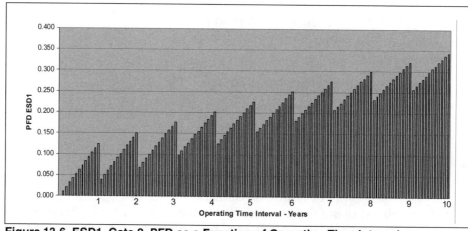

Figure 13-6. ESD1, Gate 2, PFD as a Function of Operating Time Interval

The PFD for ESD1 can be approximated with approximate simplified equations (See Chapter 4). In that case the PFD is:

$$PFD_{G2} = (\lambda_{17} + \lambda_{18} + \lambda_{19} + \lambda_{20} + \lambda_{21}) \times 8760 = 0.0674$$

for failures detected by the proof test

$$OR\ (\lambda_{33} + \lambda_{34} + \lambda_{35} + \lambda_{36} + \lambda_{37}) \times 87600 = 0.1233$$

for failures not detected by the proof test

Therefore, the approximate PFD = 0.0674 + 0.1233 = 0.191. This value is obviously pessimistic, but that is the expected result of using approximations.

ESD2 PFD Calculation

The fault tree for ESD2 is shown in Figure 13-7. ESD2 will fail dangerously if the sensor subsystem OR the logic solver subsystem OR the final element subsystem fails dangerously.

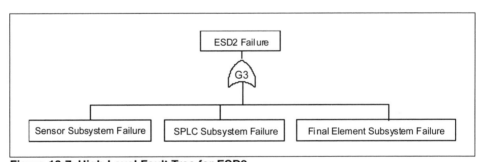

Figure 13-7. High Level Fault Tree for ESD2

A fault tree must be constructed for each subsystem in order to solve for PFD of the ESD2 portion of this SIF. Starting with the sensor subsystem, the fault tree is shown in Figure 13-8.

Gate G11 models the switch sensors located on the main shutdown valve at the inlet to the main separator (ZS subsystem). If either of these switches indicates a valve closure, the radio will send a signal to ESD2. The PFD for this branch of the fault tree can be determined gate by gate.

Gates G15 and G16 model the two individual switches. The PFD for Gate G15 is:

$$PFD_{G15} = (\lambda_{31}) \times 8760 \text{ for failures detected by the proof test}$$

$$OR\ (\lambda_{47}) \times 87600 \text{ for failures not detected by the proof test.}$$

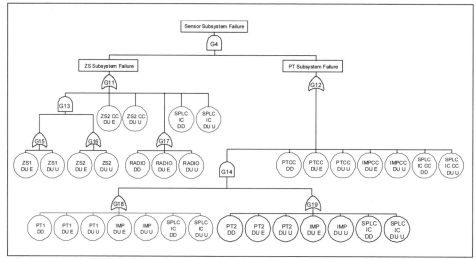

Figure 13-8. Fault Tree for ESD2 Sensor Subsystem

Gate G11 shows a failure if:

- Both switches fail dangerously
- OR switches fail dangerously common cause
- OR the radio fails dangerously
- OR the analog input circuit of the safety PLC fails dangerously.

This can be expressed with an approximate simplified equation:
$PFD_{G11} = [(\lambda_{31}) \times 8760 + (\lambda_{47}) \times 87600]^2 + \beta \times [(\lambda_{31}) \times 8760 + (\lambda_{47}) \times 87600] + (\lambda_{16}) \times 24 + (\lambda_{32}) \times 8760 + (\lambda_{48}) \times 87600 + (\lambda_{7}) \times 24 + (\lambda_{23}) \times 87600$.

The PFD_{G11} can be calculated as a function of operating time interval. A plot of this is shown in Figure 13-9.

Notice that the PFD decreases every year when the proof test is done, but it does not go to zero because the proof test is not perfect.

The sensor subsystem will also initiate a trip if either of the pressure transmitters (Gate G12) indicates an overpressure. The individual pressure transmitter failures are represented by Gates G18 and G19. G18 indicates that any dangerous failure of a pressure transmitter, its associated impulse line or the associated input circuit of the safety PLC will cause a gate failure. The PFD can be represented by approximate simplified equations. For Gate G18:

$$PFD_{G18} = (\lambda_{6}) \times 72 + (\lambda_{22}) \times 8760 + (\lambda_{38}) \times 87600 + (\lambda_{18}) \times 8760 + (\lambda_{34}) \times 87600 + (\lambda_{7}) \times 24 + (\lambda_{23}) \times 87600$$

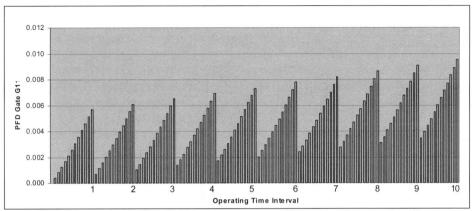

Figure 13-9. PFD of Gate G11 (ZS Subsystem) as a Function of Operating Time Interval

At Gate G14 the AND gate indicates that both transmitters must failure dangerously for gate failure. The PFD for this is:

$$PFD_{G14} = [(\lambda_6) \times 72 + (\lambda_{22}) \times 8760 + (\lambda_{38}) \times 87600 + (\lambda_{18}) \times 8760 + (\lambda_{34}) \\ \times 87600 + (\lambda_7) \times 24 + (\lambda_{23}) \times 87600]^2$$

The pressure transmitter subsystem PFD is modeled by Gate G12. Gate G12 fails if:

- both pressure transmitters fail,
- pressure transmitters fail common cause,
- impulse lines fail common cause, OR
- the safety PLC input circuits fail common cause.

This can be expressed by an approximate simplified equation:

$$PFD_{G12} = [(\lambda_6) \times 72 + (\lambda_{22}) \times 8760 + (\lambda_{38}) \times 87600 + (\lambda_{18}) \\ \times 8760 + (\lambda_{34}) \times 87600 + (\lambda_7) \times 24 + (\lambda_{23}) \times 87600]^2 \\ + \beta_{PT} \times [(\lambda_6) \times 72 + (\lambda_{22}) \times 8760 + (\lambda_{38}) \times 87600] \\ + \beta \times [(\lambda_{18}) \times 8760 + (\lambda_{34}) \times 87600] \\ + \beta \times [(\lambda_7) \times 24 + (\lambda_{23}) \times 87600]$$

The sensor subsystem, Gate 4, PFD results from an AND of Gate G11 and Gate G12.

$$PFD_{G4} = PFD_{G11} \times PFD_{G12}$$

This value can be plotted as function of operating time interval. The results are shown in Figure 13-10. Note that proof test effectiveness is much better for this subsystem.

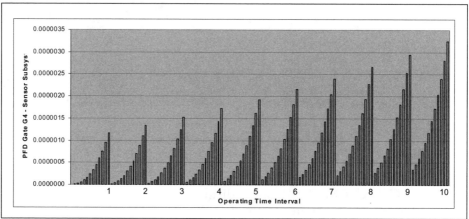

Figure 13-10. PFD Gate G4 as a Function of Operating Time Interval

The safety PLC subsystem fault tree is shown in Figure 13-11.

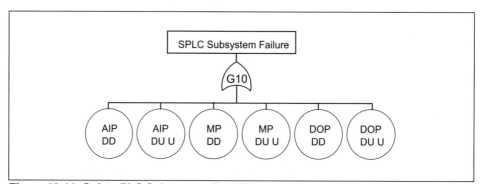

Figure 13-11. Safety PLC Subsystem Fault Tree

This fault tree states that the safety PLC will fail dangerously when the analog input module common circuitry fails dangerously, the main processor and all PLC common circuitry fail dangerously or the digital output module common circuitry fails dangerously. Note that the proof test interval for the safety PLC has been estimated at ten years. This is because the automatic diagnostics in such machines are very good and any manual proof test is not likely to find a failure that is not already detected. The manual proof test envisioned after a ten year period is a complete overhaul of the PLC with modules removed and fully tested by manufacturing test fixtures or replaced with newly manufactured units. That level of work is estimated to have 100% effectiveness.

The PFD of Gate G10 can be expressed with approximate simplified equations.

$$PFD_{G10} = (\lambda_8) \times 24 + (\lambda_{24}) \times 87600 + (\lambda_9) \times 24 + (\lambda_{25}) \times 87600 \\ + (\lambda_{10}) \times 24 + (\lambda_{26}) \times 87600.$$

NOTE: Remember that 24 hours is the specified restore time from Table 13-1.

The final element subassembly of ESD2 is represented by the fault tree of Figure 13-12.

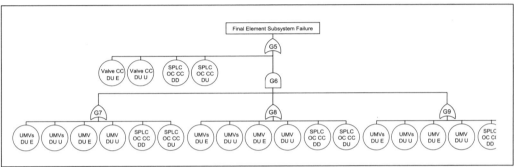

Figure 13-12. ESD2 Final Element Subsystem Fault Tree

Each valve is individually represented by Gate G7, Gate G8, and Gate G9. The 1oo3 combination is modeled by Gate G6. Common cause failures of the valves are represented by additional input to Gate G5. The PFD for Gate G5 can be approximately expressed as a simplified equation.

$$PFD_{G5} = [(\lambda_{28}) \times 8760 + (\lambda_{44}) \times 87600 + (\lambda_{29}) \times 8760 \\ + (\lambda_{45}) \times 87600 + (\lambda_{11}) \times 24 + (\lambda_{27}) \times 87600]^3 + \beta \times [(\lambda_{28}) \\ \times 8760 + (\lambda_{44}) \times 87600 + (\lambda_{29}) \times 8760 + (\lambda_{45}) \\ \times 87600 + (\lambda_{11}) \times 24 + (\lambda_{27}) \times 87600].$$

This PFD can be plotted as a function of operating time interval. That is shown in Figure 13-13.

Figure 13-13 shows the significant effect of manual proof test effectiveness. The valve and actuator proof tests are generally effective in showing problems in the actuator but poor in discovering problems due to leakage in the valve. Even a full stoke test of a valve may not show seat damage unless other observations are made. Although the overall proof test was estimated to 70 – 80% effective, the cumulative effect of PFD due to undetected failures in the manual proof test is much greater than one might estimate.

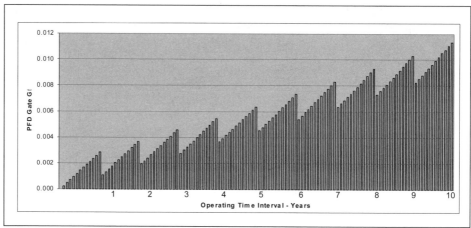

Figure 13-13. PFD Gate G5 as a Function of Operating Time Interval

To obtain the approximate PFD for ESD2, Gate 3, the PFD of each subsystem are added.

$$PFD_{G3} = PFD_{G4} + PFD_{G10} + PFD_{G5}$$

The analytical simplified equation would be complex but could be expressed.

$$\begin{aligned}
PFD_{G3} = &\{[(\lambda_{31}) \times 8760 + (\lambda_{47}) \times 87600]^2 + \beta \times [(\lambda_{31}) \times 8760 + (\lambda_{47}) \times 87600] \\
&+ (\lambda_{16}) \times 24 + (\lambda_{32}) \times 8760 + (\lambda_{48}) \times 87600 + (\lambda_7) \times 24 + (\lambda_{23}) \times 87600\} \\
&\times \{[(\lambda_6) \times 72 + (\lambda_{22}) \times 8760 + (\lambda_{38}) \times 87600 + (\lambda_{18}) \times 8760 + (\lambda_{34}) \\
&\times 87600 + (\lambda_7) \times 24 + (\lambda_{23}) \times 87600]^2 + \beta_{PT} \times [(\lambda_6) \times 72 + (\lambda_{22}) \times 8760 \\
&+ (\lambda_{38}) \times 87600] + \beta \times [(\lambda_{18}) \times 8760 + (\lambda_{34}) \times 87600] + \beta \times [(\lambda_7) \\
&\times 24 + (\lambda_{23}) \times 87600]\} + (\lambda_8) \times 24 + (\lambda_{24}) \times 87600 + (\lambda_9) \times 24 \\
&+ (\lambda_{25}) \times 87600 + (\lambda_{10}) \times 24 + (\lambda_{26}) \times 87600 + [(\lambda_{28}) \times 8760 \\
&+ (\lambda_{44}) \times 87600 + (\lambda_{29}) \times 8760 + (\lambda_{45}) \times 87600 + (\lambda_{11}) \times 24 \\
&+ (\lambda_{27}) \times 87600]^3 + \beta \times [(\lambda_{28}) \times 8760 + (\lambda_{44}) \times 87600 + (\lambda_{29}) \times 8760 \\
&+ (\lambda_{45}) \times 87600 + (\lambda_{11}) \times 24 + (\lambda_{27}) \times 87600]
\end{aligned}$$

A close inspection will show that the terms equal the specific terms from each gate.

When the PFD of Gate G3 is plotted as a function of operating time interval, Figure 13-14 results.

Overall SIF PFD and PFDavg Calculation

The overall SIF PFD can be obtained using Gate G1. A plot of Gate PFD as a function of operating time interval is shown in Figure 13-15.

Oil and Gas Production Facilities   207

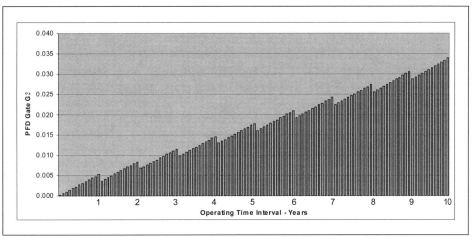

Figure 13-14. PFD Gate G3 as a Function of Operating Time Interval

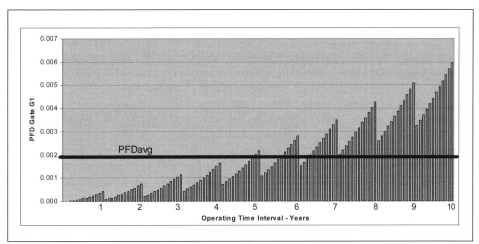

Figure 13-15. PFD Gate G1 as a Function of Operating Time Interval

An analytical expression for PFD can be done.

$$\text{PFD}_{G1} = \text{PFD}_{G2} + \text{PFD}_{G3}$$

The entire expression is:

$$\text{PFD}_{G1} = (\lambda_{17} + \lambda_{18} + \lambda_{19} + \lambda_{20} + \lambda_{21}) \times 8760 + (\lambda_{33} + \lambda_{34} + \lambda_{35} + \lambda_{36} + \lambda_{37}) \times 87600$$

$$+ \{[(\lambda_{31}) \times 8760 + (\lambda_{47}) \times 87600]^2 + \beta \times [(\lambda_{31}) \times 8760 + (\lambda_{47}) \times 87600]$$
$$+ (\lambda_{16}) \times 24 + (\lambda_{32}) \times 8760 + (\lambda_{48}) \times 87600 + (\lambda_{7}) \times 24 + (\lambda_{23}) \times 87600\}$$
$$\times \{[(\lambda_{6}) \times 72 + (\lambda_{22}) \times 8760 + (\lambda_{38}) \times 87600 + (\lambda_{18}) \times 8760 + (\lambda_{34}) \times 87600$$
$$+ (\lambda_{7}) \times 24 + (\lambda_{23}) \times 87600]^2 + \beta_{PT} \times [(\lambda_{6}) \times 72 + (\lambda_{22}) \times 8760$$
$$+ (\lambda_{38}) \times 87600] + \beta \times [(\lambda_{18}) \times 8760 + (\lambda_{34}) \times 87600]$$

$$+ \beta \times [(\lambda_7) \times 24 + (\lambda_{23}) \times 87600]\}$$
$$+ (\lambda_8) \times 24 + (\lambda_{24}) \times 87600 + (\lambda_9) \times 24 + (\lambda_{25}) \times 87600$$
$$+ (\lambda_{10}) \times 24 + (\lambda_{26}) \times 87600$$
$$+ [(\lambda_{28}) \times 8760 + (\lambda_{44}) \times 87600 + (\lambda_{29}) \times 8760 + (\lambda_{45}) \times 87600$$
$$+ (\lambda_{11}) \times 24 + (\lambda_{27}) \times 87600]^3 + \beta \times [(\lambda_{28}) \times 8760 + (\lambda_{44}) \times 87600$$
$$+ (\lambda_{29}) \times 8760 + (\lambda_{45}) \times 87600 + (\lambda_{11}) \times 24 + (\lambda_{27}) \times 87600]$$

The PFDavg for entire safety instrumented function can be calculated a number of different ways. One approach is to calculate an arithmetic average of the PFD as a function of operating time interval. When this is done on a spreadsheet, the answer is approximately 0.0019. Per the low demand mode PFDavg chart, this qualifies for SIL2.

Analysis of PFD Results

Comparing the PFD from the two ESD systems (Table 13-3) it can be seen that ESD2 provides an order of magnitude better performance. While ESD1 does help somewhat, it is likely that the false trip rate contribution from this equipment will negate its overall life cycle cost benefit. Consideration should be given to simply eliminating ESD1.

Table 13-3. PFD Comparison of the Two ESD Systems

Overview	Gate	PFD
ESD1 PFD	G2	0.343287
ESD2 PFD	G3	0.034697

A stacked bar chart of PFD contribution to ESD2 is shown in Figure 13-16. A detailed analysis of the PFD contribution from each subsystem in ESD2 shows that the logic solver contributes the most to the PFD. Given this situation, better technology (a SIL3 rated safety PLC) could improve the PFDavg if necessary. Alternatively, a manual proof test procedure could be developed to detect a portion of the failures each year.

The sensor subsystem contributes an insignificant amount to the PFD of ESD2. That probably represents an opportunity to reduce the quantity of equipment and also reduce the lifecycle cost of manual proof testing. Comparing the failure rates of the switches that indicate closure of the valve to the failure rate of the radio system, it is clear that one of the two switches could easily be eliminated. This will reduce the false trip rate as well as lower capital and lifecycle cost. One could go further, however, by looking at the safety contribution of each of the sensor types. The PFD contribution of the ZS sensor subsystem (Gate G11) and the PT sensor subsystem (Gate G12) is shown in Table 13-4.

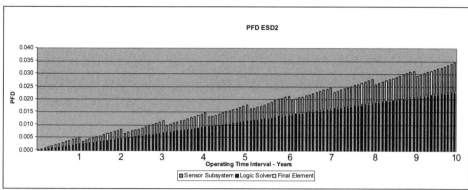

Figure 13-16. ESD2 - PFD Contribution of Major Subsystems

Table 13-4. PFD Contribution of Sensor Subsystems

Subsystem	PFD
ZS sensor subsystem	0.00954600
PT sensor subsystem	0.00033955

Given the high PFD of the ZS sensor subsystem, it is likely that the entire subsystem could be eliminated without a significant penalty in overall PFD. This would again reduce the false trip rate and lifecycle cost. Of course, the calculations must be repeated for each proposed design to assure that the changes will not impact safety integrity below the required levels.

Alternative SIF Designs

Considering the results of the SIF verification analysis it appears that an alternative design could not only save capital expense but could provide higher safety integrity and a lower false trip rate. The proposal for this design is shown in Figure 13-17. The ZS sensor subsystem is eliminated and the ESD1 subsystem is eliminated.

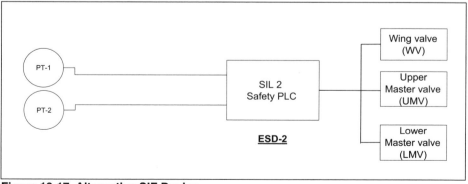

Figure 13-17. Alternative SIF Design

Eliminating the ZS sensor subsystem and the ESD1 subsystem without improving the design elsewhere will result in a higher PFDavg. In order to find out how much, the PFDavg will be calculated. The top level of the fault tree for the proposed design is simplified and shown in Figure 13-18, which maintains the same gate numbers for comparison to the previous fault trees.

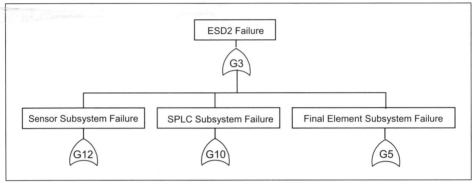

Figure 13-18. Fault Tree of Simplified Design

The PFD of this fault tree can be calculated from a spreadsheet (see Appendix C for different methods). The PFDavg is 0.017. The result for the same SIF design and the same input parameters obtained using the exida SILver^{TM} tool is also 0.017. This is shown in Figure 13-19. Unfortunately, this value of 0.017 has dropped below the SIL 2 level, so improvements must be made.

Safety Instrumented Function Performance Metrics for SIF High Pressure	
Average Probability of Failure on Demand (PFDavg)	1.74E-02
Safety Integrity Level (PFDavg)	1
Safety Integrity Level (Architectural Constraints IEC 61508)	1
Risk Reduction Factor	57
MTTFS (years)	3.78

Figure 13-19. SILver screen for simplified design SIF

An improvement in safety integrity can be expected if a higher integrity safety PLC were used. Given the improvement, perhaps a SIL 2 could be achieved for the SIF. In addition, diagnostic comparison between the two pressure transmitters could be done in the logic solver. This would result in an improvement also. The calculation was done again for the improved design using the exida SILver™ tool. The result using a SIL 3 logic solver is 0.00576. The details are shown in Figure 13-20.

This design meets SIL 2 with a lower false trip rate and lower life cycle cost.

PFDavg Calculations for all wells

Since there are 5 separate wells connected to the high pressure separator, the closing of the inlet valve to the separator requires that all the wells shutdown. If one well fails to shutdown then the line pressure may still rise creating a hazardous event.

A separate Safety Instrumented Function that needs to be analyzed is the closing of all wells when the main inlet separator valve closes. The question is "What is the probability of any one well failing to trip when the main inlet separator valve closes?"

A fault tree for this SIF is shown in Figure 13-21 and is based on the probability of any one of the five trip systems failing dangerously.

A Solution Using the OR Gate Approach

The probability of any one well failing on demand has been calculated. If five are present and any one can cause system failure then the approximate average probability of failure on demand can be obtained by adding the PFDavg of each well.

$$PFDavg_{(5\ wells)} = PFDavg(single) + PFDavg(single) + PFDavg(single) + PFDavg(single) + PFDavg(single)$$

Using the single well result from the final alternative design of PFDavg = 0.00576:

$$PFDavg_{(5\ wells)} = 5 \times 0.00576 = 0.0288$$

Note: This is an approximation.

A Solution Using the AND Gate Approach

The probability of a well failing to trip when the main inlet separator valve closes is (1- the probability of all five wells tripping when the main inlet separator valve closes).

Since we know the probability of a single well failing to trip when the main inlet separator valve closes, then the probability of a single well

Safety Instrumented Function Performance Metrics for SIF High Pressure	
Average Probability of Failure on Demand (PFDavg)	5.76E-03
Safety Integrity Level (PFDavg)	2
Safety Integrity Level (Architectural Constraints IEC 61508)	3
Risk Reduction Factor	173
MTTFS (years)	7.64

Sensor Part Information	
Sensor Group(s)	Edit
(1) Pressure Transmitters	Details
PFDavg Sensor Part	7.83E-06
MTTFS Sensor Part (years)	384.51
Maximum SIL allowed (Architectural Constraints IEC 61508)	3

Logic Solver Part Information	
Logic Solver	Edit
(1) SIL 3 Logic Solver	Details
PFDavg Logic Solver Part	3.78E-04
MTTFS Logic Solver Part (years)	194.92
Maximum SIL allowed (Architectural Constraints IEC 61508)	3

Final Element Part Information	
Final Element Group(s)	Edit
(1) V	Details
PFDavg Final Element Part	5.38E-03
MTTFS Final Element Part (years)	8.12
Maximum SIL allowed (Architectural Constraints IEC 61508)	4

Figure 13-20. Simplified Design with SIL 3 Logic Solver and Transmitter Comparison

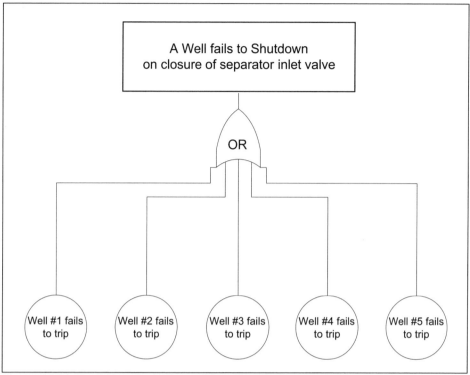

Figure 13-21. Fault Tree for Failure of Any One Well

tripping when the main inlet separator valve closes is (1-PFDavg of well #1 failing to trip on high pressure)

$$= 1 - 0.00576$$

$$= 0.99424$$

The probability of all five wells tripping when the main inlet separator valve

$$= (0.99424)^5$$

$$= 0.97153$$

Therefore the probability of a well failing to trip when the main inlet separator valve closes is

$$= 1 - 0.97153$$

$$= 0.0285$$

Exercises

13-1. Complete the conceptual design requirements for a SIL 3 High Integrity Pressure Protection System (HIPPS) required to protect a main separator fed from multiple well heads from over pressuring. A simplified P&ID for the installation is shown in Figure 13-22.

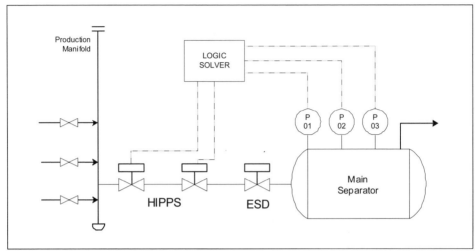

Figure 13-22. Simplified P&ID

13-2. Based on the assumptions and conceptual design, verify that the design proposed will satisfy the SIL requirements.

REFERENCES AND BIBLIOGRAPHY

1. API 14C, *Recommended Practice for Analysis, Design, Installation and Testing of Basic Surface Safety Systems for Offshore Production Platforms*, Seventh Edition, D.C.: Washington, American Petroleum Institute, March 2001.

2. IEC 61508, *Functional Safety of Electrical / Electronic / Programmable Electronic Safety-related Systems*, IEC, 2000.

3. ANSI/ISA-84.00.01-2004, *Functional Safety: Safety Instrumented Systems for the Process Industry Sector – Parts 1, 2, and 3 (IEC 61511 Mod)*. ISA, 2004.

4. *Safety Equipment Reliability Handbook*. exida, 2003. (available from ISA)

14
Chemical Industry

Introduction

In this chapter two separate examples commonly found in chemical plants will be presented. The examples are intended to demonstrate the importance of properly selecting which equipment needs to be included in a SIF, and how equipment selection affects SIL values achieved.

Note that the example SIL levels provided in this chapter are only examples. They are not to be assumed recommended levels of protection. The selection of an appropriate Safety Integrity Level (SIL) is site-specific and the analysis requires selecting criteria for tolerable risk, and evaluating process conditions, specific chemicals, equipment design-limits, control schemes, process conditions, and unique hazards. Experts in process engineering, instrumentation, operations, and process safety should undertake SIL selection.

Reactor

Reactors are used in chemical plants to produce a wide variety of products. Many of the reactors are exothermic and the temperatures and pressures in the reactors can rise considerably if a reaction upset occurs. The need for an SIS has been identified for the reactor system shown in the simplified drawing of Figure 14-1.

The SIF description for the reactor is shown in Table 14-1. The SIF identified is a function that closes the feed valve XV-01 when either temperature or pressure gets above limit.

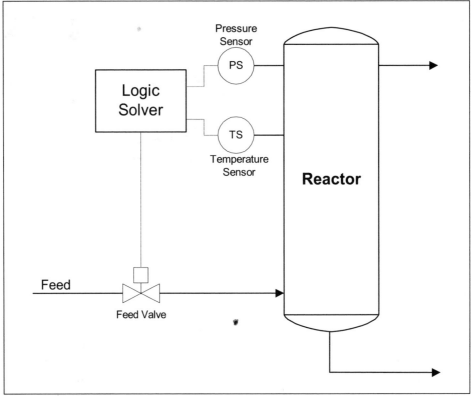

Figure 14-1. Reactor

Table 14-1. SIF for Reactor

SIF	SIF INITIATORS	HAZARDOUS EVENT	INPUT SIGNALS	SIL	PRIMARY OUTPUTS	AUXILIARY OUTPUTS
14.1	High temperature and/or pressure in the reactor caused by blockage of reactor outlet line.	Rupture of vessel causing a release of flammable gas to atmosphere resulting in explosion and equipment damage.	PS-01 and TS-02	2	XV-01 Closes	None

The process hazards analysis for this vessel stated that either high temperature or high pressure is an indication of a potential runaway reaction that may result in an explosion. The safety requirements specification states that the SIF shall be SIL 2 and a nuisance or spurious trip should be better than one in every five years. The process will be operated for mission time of ten years after which a major overhaul and rebuild will take place.

Note: The selection of both temperature and pressure as SIF inputs (1oo2 voting) has to be carefully reviewed and analyzed. The basis for this selection must determine a very close correlation between temperature and pressure, i.e. when a demand occurs, both temperature and pressure will exceed prescribed limits.

The SIF components with the associated data are available as shown in Table 14-2.

Table 14-2. Instrument Failure Rate and Failure Mode Data

	Lambda	%Safe	Automatic Test Diagnostic Coverage CD	Repair Time (RT)	Proof Test Coverage	Lambda S		Lambda DD		Lambda DU E		Lambda DU U		Reference
Pressure transmitter	1.50E-06	10%	56%	72	95%	(1 =	1.50E-07	(15 =	7.50E-07	(29 =	5.70E-07	(43 =	3.00E-08	Fig. 8-3
Pressure switch	6.00E-06	40%	0%	72	95%	(2 =	2.40E-06	(16 =	0.00E+00	(30=	3.42E-06	(44 =	1.80E-07	Fig. 8-2
Impulse line	2.50E-07	0%	0%	72	95%	(3 =	0.00E+00	(17 =	0.00E+00	(31 =	2.38E-07	(45 =	1.25E-08	EXI-SILver LOW
Temperature switch	6.00E-06	40%	0%	72	95%	(4 =	2.40E-06	(18 =	0.00E+00	(32=	3.42E-06	(46=	1.80E-07	[EXI03] pg. 78
Temperature transmitter (Thermocouple)	6.50E-06	2%	87%	72	95%	(5 =	1.50E-07	(19 =	5.50E-06	(33=	8.08E-07	(47=	4.25E-08	[EXI03] pg. 79
Safety PLC Digital Input Channel (IC)	2.02E-07	65%	94%	24	99%	(6 =	1.31E-07	(20 =	6.70E-08	(34 =	3.96E-09	(48 =	4.00E-11	[EXI03] pg.124
Safety PLC Digital Input Processing (IP)	1.00E-06	60%	95%	24	99%	(7 =	6.00E-07	(21 =	3.80E-07	(35 =	1.98E-08	(49 =	2.00E-10	[EXI03] pg.124
Safety PLC Analog Input Channel (IC)	1.02E-07	50%	94%	24	99%	(8 =	5.10E-08	(22 =	4.80E-08	(36 =	2.97E-09	(50 =	3.00E-11	[EXI03] pg.124
Safety PLC Analog Input Processing (IP)	2.00E-06	50%	90%	24	99%	(9 =	1.00E-06	(23 =	9.00E-07	(37 =	9.90E-08	(51 =	1.00E-09	[EXI03] pg.124
Safety PLC Main Processing (MP)	1.25E-05	74%	95%	24	99%	(10 =	9.25E-06	(24 =	3.10E-06	(38 =	1.49E-07	(52 =	1.50E-09	[EXI03] pg.124
Safety PLC Digital Output Processing (OP)	1.00E-06	80%	95%	24	99%	(11 =	8.00E-07	(25 =	1.90E-07	(39 =	9.90E-09	(53 =	1.00E-10	[EXI03] pg.124
Safety PLC Digital Output Channel (OC)	1.20E-07	75%	99%	24	99%	(12 =	9.00E-08	(26 =	2.97E-08	(40 =	2.97E-10	(54 =	3.00E-12	[EXI03] pg.124
Solenoid valve	6.00E-06	60%	0%	168	70%	(13 =	3.60E-06	(27=	0.00E+00	(41 =	1.68E-06	(55 =	7.20E-07	[EXI03] pg.155
Ball Valve	2.45E-06	67%	0%	168	70%	(14=	1.65E-06	(28 =	0.00E+00	(42 =	5.60E-07	(56 =	2.40E-07	EXI-SILver

For each component, a set of failure rate and related parameters are given. The total failure rate, Lambda, is given as well as the percentage of the total failure rate that is safe. The diagnostic coverage factor for any automatic diagnostics within the SIF is estimated or obtained from references. A restore time is estimated for each component. A "manual proof test effectiveness" (diagnostic coverage of the manual proof test) is also estimated. Using these numbers, three dangerous failure rates are calculated and one safe failure rate is calculated. For the safety PLC, a one-year proof test will be done using built-in test routines that are rated 99% effective.

A conceptual design for the lowest-cost, lowest-complexity system that meets the requirements must now be done. From the situation description the simplest architecture for the safety function is shown in Figure 14-2.

To verify that the design meets the requirements, a probabilistic calculation must be done to determine PFDavg, MTTFS and Architecture

218   Chemical Industry

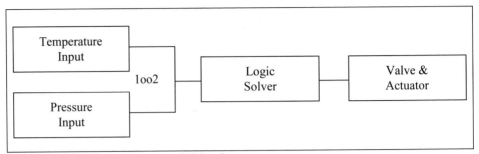

Figure 14-2. Proposed Reactor SIF Design

Constraints. The first design proposal uses a pressure switch and temperature switch as the sensing devices. A fault tree for the PFDavg for this design is shown in Figure 14-3.

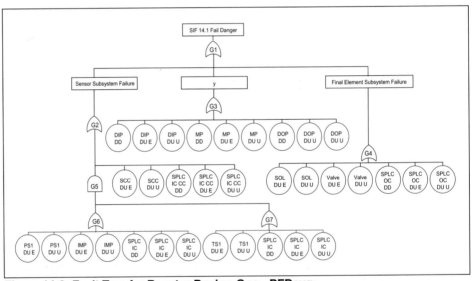

Figure 14-3. Fault Tree for Reactor Design One - PFDavg

The PFDavg can be obtained by solving the fault tree for PFD as a function of operating time and averaging the time dependent results (See Appendix C). The PFD can be solved gate by gate using simplified approximation equations. These equations use simplified approximations where the 8760 hour time period is the one year test interval, the 87600 hour time period is the ten year mission time and the 24 hour time period is the restore time for the dangerous detected failures in the safety PLC.

Gate 6 accounts for the pressure switch, the impulse line and the connected safety PLC digital input circuit. The simplified approximation for gate 6, PFD is:

$$PFD_{G6} = (\lambda_{30}) \times 8760 + (\lambda_{44}) \times 87600 + (\lambda_{31}) \times 8760 + (\lambda_{45}) \times 87600 \\ + (\lambda_{20}) \times 24 + (\lambda_{34}) \times 8760 + (\lambda_{48}) \times 87600$$

Gate 7 accounts for the temperature switch and the safety PLC digital input circuit to which it is connected. The simplified approximation for gate 7, PFD is:

$$PFD_{G7} = (\lambda_{32}) \times 8760 + (\lambda_{46}) \times 87600 + (\lambda_{20}) \times 24 + (\lambda_{34}) \times 8760 + (\lambda_{48}) \times 87600$$

Gate 5 represents the 1oo2 vote of the pressure switch and the temperature switch. The simplified approximation for gate 5, PFD is:

$$PFD_{G5} = [(\lambda_{30}) \times 8760 + (\lambda_{44}) \times 87600 + (\lambda_{31}) \times 8760 + (\lambda_{45}) \times 87600 \\ + (\lambda_{20}) \times 24 + (\lambda_{34}) \times 8760 + (\lambda_{48}) \times 87600] \times [(\lambda_{32}) \times 8760 + (\lambda_{46}) \times 87600 \\ + (\lambda_{20}) \times 24 + (\lambda_{34}) \times 8760 + (\lambda_{48}) \times 87600]$$

Gate 2 represents the entire sensor subsystem. In addition to the Gate 5 results, common cause failures of the two sensors and the safety PLC input circuits are included. Given that the two switches are likely to be similar technology, a common cause beta factor of 5% was chosen. The simplified approximation equation for gate 2, PFD is:

$$PFD_{G2} = [(\lambda_{30}) \times 8760 + (\lambda_{44}) \times 87600 + (\lambda_{31}) \times 8760 + (\lambda_{45}) \times 87600 \\ + (\lambda_{20}) \times 24 + (\lambda_{34}) \times 8760 + (\lambda_{48}) \times 87600] \times [(\lambda_{32}) \times 8760 + (\lambda_{46}) \\ \times 87600 + (\lambda_{20}) \times 24 + (\lambda_{34}) \times 8760 + (\lambda_{48}) \times 87600] \\ + \beta \times [(\lambda_{30} + \lambda_{32}) / 2 \times 8760 + (\lambda_{44} + \lambda_{46}) / 2 \times 87600 \\ + (\lambda_{20}) \times 24 + (\lambda_{34}) \times 8760 + (\lambda_{48}) \times 87600]$$

The PFD calculated value is 0.0045. When the time dependent PFD is averaged over the operating time interval the PFDavg is 0.0019.

Gate 3 accounts for the circuitry used within the safety PLC. One digital input module, one digital output module and all common circuitry (Main Processor) are included. The simplified approximation equation for gate 3, PFD is:

$$PFD_{G3} = (\lambda_{21}) \times 24 + (\lambda_{35}) \times 8760 + (\lambda_{49}) \times 87600 + (\lambda_{24}) \times 24 \\ + (\lambda_{38}) \times 8760 + (\lambda_{52}) \times 87600 + (\lambda_{25}) \times 24 + (\lambda_{39}) \times 8760 + (\lambda_{53}) \times 87600$$

The PFD calculated value is 0.0018. The average of time dependent PFD, PFDavg, is 0.001.

Gate 4 represents the final element subsystem. Included in this portion of the fault tree is the solenoid and actuator/valve. The simplified approximation equation for gate 4, PFD is:

$$PFD_{G4} = (\lambda_{41}) \times 8760 + (\lambda_{55}) \times 87600 + (\lambda_{42}) \times 8760 + (\lambda_{56}) \\ \times 87600 + (\lambda_{26}) \times 24 + (\lambda_{40}) \times 8760 + (\lambda_{54}) \times 87600$$

The PFD value is 0.1037 and the PFDavg value is 0.0526.

Table 14-3 shows the PFD and PFDavg values for the individual subsystems.

Table 14-3. Reactor Design 1, PFD and PFDavg Subsystem Values

		PFD	PFDavg
Sensor subsystem	G2	0.0045	0.0019
Safety PLC subsystem	G3	0.0018	0.0010
Final Element subsystem	G4	0.1037	0.0526

The PFD for the entire SIF is approximated by summing the PFD of gate 2, gate 3, and gate 4. The PFD as a function of operating time interval is shown in Figure 14-4.

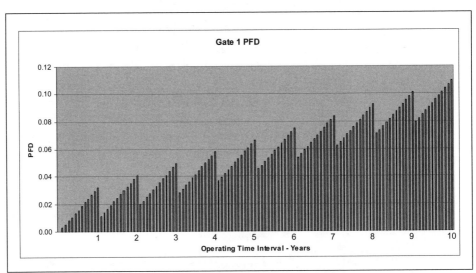

Figure 14-4. SIF PFD as a Function of Operating Time Interval

The PFDavg is calculated taking the average of the PFD. In this case the result is 0.055.

Figure 14-5 shows the results for the same problem obtained with the SILver™ tool. These results are slightly better because the simplified equation approximations are not used in the tool.

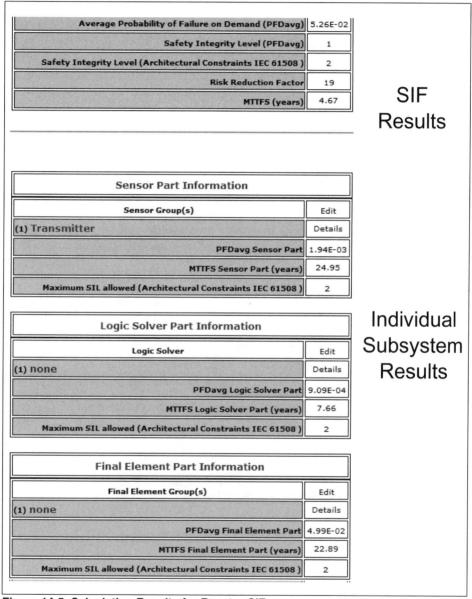

Figure 14-5. Calculation Results for Reactor SIF

In any case, the design does not meet the requirements of SIL 2 for PFDavg. It can be clearly seen from Table 14-3 or Figure 14-5 that the problem is the final element subsystem. Consequently this part of the safety function needs to be changed in order to meet the SIL 2 requirement. In order to increase the safety integrity of the final element subsystem, a 1oo2 voting configuration could be considered. A common cause beta factor of 10% will be estimated based on identical valves in identical service.

The results were recalculated with two valves piped in a 1oo2 voting scheme and are shown in Figure 14-6. Although there is a significant improvement, the total result of 0.0109 still does not quite meet SIL 2! This is still primarily due to the high PFDavg value or final elements.

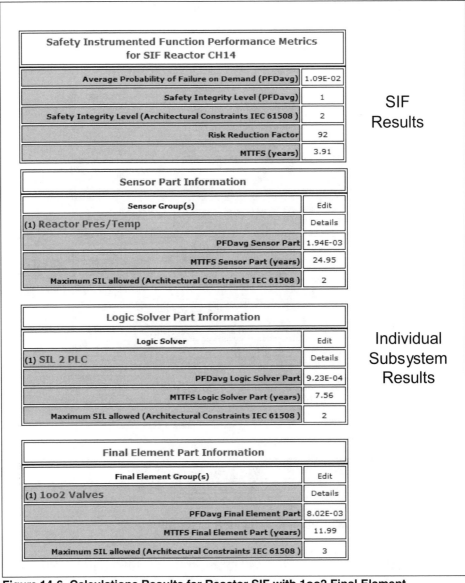

Figure 14-6. Calculations Results for Reactor SIF with 1oo2 Final Element

Another alternative design might utilize one valve with partial valve stroke testing. As can be seen in Chapter 11, the effectiveness of a partial valve stroke test depends on the type of valve and the service conditions. Figure 14-7 shows the particular data from an actuator/valve FMEDA for the reactor example.

Component	Final Element Data						
	Failure rates (1/hr)					Architectural Constraint Type	Safe Failure Fraction
	Dangerous Detected	Dangerous Undetected	Safe Detected	Safe Undetected	No Effect		
leg 1 Interface: Fisher Controls DVC6000 0-20mA PVST data from database		3.60E-08 3.60E-08	3.90E-08 3.90E-08	6.50E-07 6.50E-07		A	95.0 95.0
leg 1 Actuator-Valve Combination: Generic Air Operated Ball Valve, Hard Seat - Clean service, Full Stroke, Close To Trip	4.80E-07	8.00E-07 3.20E-07	1.65E-06	1.65E-06		A	67.3 86.9

Figure 14-7. Partial Valve Stroke Data from FMEDA

A design using a single valve with partial valve stroke testing would meet SIL 2 per the results shown in Figure 14-8. In addition, the MTTFS also meets the minimum criteria of five years. Of course there may be many other designs that meet both the SIL 2 integrity criteria and the minimum MTTFS of five years.

Separator Vessel

Separation vessels can be found in most chemical plants and are used to separate inlet streams into different phases. See Figure 14-9. In the chemical industries it is common to find inlet streams consisting of a liquid with entrained gas at a high pressure. As the multiphase stream enters the separator the temperature and pressure decreases. This causes the gas and liquid to separate. The liquid level in the separator is normally controlled by a level control loop in the BPCS and is normally continuously pumped to another vessel for further processing. The gas exiting the top of the vessel is normally compressed and distributed.

LIC-01 controls the level in the vessel as part of the BPCS system. SIF-15 is the safety instrumented function. The input signal to the SIF is LT-2, and the output signals from the logic solver operate XV-1, XV-2, and P-01.

The cause and effects diagram associated with the above SIF will show that on low low level sensed by LT-2, pump P1 will stop and ESD valves XV-01 and XV-02 will close.

Safety Instrumented Function Performance Metrics for SIF Pressure Loop

Average Probability of Failure on Demand (PFDavg)	8.65E-03
Safety Integrity Level (PFDavg)	2
Safety Integrity Level (Architectural Constraints IEC 61508)	2
Risk Reduction Factor	116
MTTFS (years)	5.24

Sensor Part Information

Sensor Group(s)	Edit
(1) Transmitter	Details
PFDavg Sensor Part	1.94E-03
MTTFS Sensor Part (years)	24.95
Maximum SIL allowed (Architectural Constraints IEC 61508)	2

Logic Solver Part Information

Logic Solver	Edit
(1) none	Details
PFDavg Logic Solver Part	9.09E-04
MTTFS Logic Solver Part (years)	7.66
Maximum SIL allowed (Architectural Constraints IEC 61508)	2

Final Element Part Information

Final Element Group(s)	Edit
(1) 1oo1 Partial Valve Stroke	Details
PFDavg Final Element Part	5.82E-03
MTTFS Final Element Part (years)	49.09
Maximum SIL allowed (Architectural Constraints IEC 61508)	2

Figure 14-8. Calculation Results for Reactor SIF Using 1oo1 Final Element with Partial Valve Stroke Testing

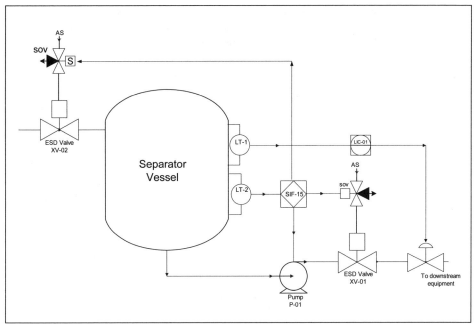
Figure 14-9. Separator Vessel

Potential hazardous events associated with a loss of level in the vessel can be:

1. Gas Blowby – High pressure gas can flow into the pumping system with severe consequences. Examples are:
   - Damage to pump
   - Release of flammable or toxic gases to the atmosphere
   - Fire
   - Damage to the piping system
   - Personnel injuries and/or fatalities
2. Loss of liquid flow to pump causing pump damage resulting in gas release, fire, and potential fatalities

As part of the SIL selection, more than one safety instrumented function may have to be defined because there are different hazardous events associated with the loss of level. One SIF can be associated with the loss of the pump and another SIF can be associated with the effects on downstream equipment. Since the consequences, likelihood, and safeguards are different for the various hazardous events, the SIL determined can be different for each SIF. The SRS for each SIF must clearly state the hazardous event being mitigated or prevented by each individual SIF.

For each SIF defined, the primary versus auxiliary equipment must be clearly designated. Table 14-4 shows the three different SIF outputs that are possible parts of one SIF.

The primary SIF outputs are those that if operated successfully will prevent the hazardous event. The auxiliary outputs are outputs that are part of the SIF action but are not required to prevent the hazardous event from occurring.

Table 14-4. Alternative Output Selections

SIF	SIF INITIATORS	HAZARDOUS EVENT	INPUT SIGNALS	SIL	SIF OUTPUTS	AUXILIARY OUTPUTS
14.1 Case 1	Loss of level in vessel may result in downstream equipment being over pressurized resulting in line rupture	Release of flammable gas to atmosphere resulting in fire, equipment damage and 2 possible fatalities	LT-02 (1oo1)	1	XV-01 Closes	P-01 Stops. XV-02 closes
14.1 Case 2	Loss of level in vessel may result in downstream equipment being over pressurized resulting in line rupture	Release of flammable gas to atmosphere resulting in fire, equipment damage and 2 possible fatalities	LT-02 (1oo1)	1	XV-01 Closes OR XV-01 Closes	P-01 Stops.
14.1 Case 3	Loss of level in vessel may result in downstream equipment being over pressurized resulting in line rupture	Release of flammable gas to atmosphere resulting in fire, equipment damage and 2 possible fatalities	LT-02 (1oo1)	1	XV-01 Closes AND XV-01 Closes AND P-01 Stops	

Will the hazard be prevented if only XV-01 closes as described in case one?

If so, that could be a valid SIF description. If however the hazard would be prevented if either XV-01 OR XV-02 closes, then case two is a much better description. If all three final elements are required to operate then case three is the best description. The results of the SIL verification calculations for case 1, 2, and 3 using failure rate data for generic components are shown in Table 14-5.

The following assumptions were made to determine the results:

- Plant Startup Time – 24 hours
- Maximum time between plant turnarounds – 5 years
- MTTR – 8 hours
- Proof Test Interval – 1 year
- Proof Test Coverage – 90%

- Level transmitter – Generic level transmitter
- Logic Solver – General purpose PLC
- Final Elements – 3-way generic solenoid with generic air operated ball valve with hard seats

Table 14-5. Calculation Results for Different Output Selections

SIF	PFDavg Sensor	PFDavg Logic solver	PFDavg Final Elements	PFDavg System	RRF	MTTFS	SIL achieved	Notes
14.1 Case 1	7.61E-03	1.26E-02	1.93E-02	3.86E-02	26	6.33	1	1oo1 voting
14.1 Case 2	7.61E-03	1.26E-02	5.70E-04	2.07E-02	48	4.9	1	1oo2 voting for valves
14.1 Case 3	7.61E-03	1.26E-02	4.06E-02	6.03E-02	17	8.72	1	3oo3 voting for final elements

The above analysis shows the importance of clearly defining SIF outputs. Of the three cases, case two provides the best safety performance but the worst nuisance trip rate.

For case three, a risk reduction factor of 17 is obtained. This is almost SIL 1 borderline. It is the final element configuration that causes this reduction in RRF.

It is therefore essential that the inputs and outputs associated with all Safety Instrumented Functions (SIF) be properly defined to address the specific hazardous event being prevented or mitigated by the SIF. Failing to do so will result in erroneous SIL verification results.

Exercise

14-1. Calculate the PFDavg for the final element portion of a SIF function consisting of 2 trip valves with 1oo2 voting. Each valve is solenoid operated from 2 solenoid valves with 2oo2 voting. Refer to the sketch and data below.

Reliability Data Final Element components

Component	Failure Rates [1/h]					Arch. Type
	DD	DU	SD	SU	No Effect	
Generic 2oo2 solenoid configuration	-	2.65E-07	-	7.13E-06	-	A
Generic Air Operated Ball Valve, Hard Seat, Full Stroke	-	8.00E-07		1.65E-06	-	A

Other data:

MTTR:	8 hours
Unit startup time:	24 hours
Test Interval:	12 months
Proof test coverage:	80%
Beta:	10%
Lifetime of plant:	25 years

REFERENCES AND BIBLIOGRAPHY

1. *Safety Equipment Reliability Handbook*, exida, 2003. (available from ISA)

15
Combined BPCS/SIS Designs

Introduction

Many SIF designs have been done using the BPCS to implement the safety functions over the years. This practice is now banned in some functional safety standards and strongly discouraged in other standards. The ANSI/ISA-84.00.01-2004 (IEC 61511 Mod) standard (Ref. 1) does however allow the use of combined BPCS and SIS designs under certain conditions and when justified by detailed analysis. In general, the analysis work can be extensive and usually results in complete separation between the BPCS and the SIS. However, when there is a strong justification, the analysis may be worth the effort in the quest for an optimal design.

Analysis Tasks

Any engineer who wants to attempt combining control and safety must go through a series of analysis and classification tasks that should spell out the potential problems and identify any design flaws. The steps include:

1. Classification of all equipment used in the SIS,

2. Justification of all equipment used in the SIS,

3. Analysis of possible "initiating events" caused by failure of equipment used in the combined SIS and control system and comparison of resulting hazard rate to tolerable limits,

4. Practicality analysis of resulting operation and maintenance procedures.

When these steps are done, alternative designs that meet the requirements of the standard can be compared to determine the optimal solution.

Equipment Classification

All equipment used in the SIS must be classified as a safety instrumented system. The design, installation, operation and maintenance process must follow all the rules of ANSI/ISA-84.00.01-2004 (IEC 61511 Mod), put there to prevent systematic faults. If this is not done, the standard clearly states that any safety instrumented function cannot have a risk reduction greater than 10. This is the bottom of SIL1 range so, in effect, that design cannot meet SIL 1 requirements. The practical effect of this requirement is that a designer cannot combine control functions and safety functions in the same equipment unless the equipment is classified as a safety instrumented system and follows all the design rules of the standard.

Equipment Justification

If the equipment is classified as a SIS, then the next task is equipment justification. This must be done by either a "prior use" justification or equipment must be assessed per IEC 61508 (Ref. 2). Most choose the IEC 61508 certification route, especially for the logic solver, since the documentation burden is high and data gathering can be painful and fruitless when good systems are not in place. This is discussed in detail in Chapter 7.

Initiating Event Analysis

The next task is an analysis showing that no control system failure can cause an initiating event that can result in a hazard. If control system failure can initiate a hazardous sequence, then safety instrumented functions MUST NOT be designed into common equipment without detailed quantitative risk analysis. That language in the standard is strong and clear. Most of the time, initiating event analysis shows a problem with combined control and safety.

If control system equipment failure can cause an initiating event, then quantitative analysis must be done for all components where failure might initiate a hazard. For those failures with no other protection layer, the frequency of failure will result directly in an incident. The detailed quantitative analysis must show that these failures will not increase risk beyond tolerable levels.

Practicality Analysis of Operation and Maintenance

Assuming that all that analysis work is done and the designer still meets tolerable risk requirements with a combined control system and SIS, then the designer should consider the maintenance and operational procedures required for the SIS. The standard has many requirements for management of change, proof testing and security. A designer should consider the effects of having all these procedures in effect for the control system. Will the system owner really follow all these procedures? If the procedures are followed will the lifecycle cost be significantly increased?

Alternative Designs

Figure 15-1 shows a set of equipment. Transmitter T2 is connected directly to the SIS, and transmitter T1 is the sensor for the BPCS control loop. The SIS logic solver is separate and independent from the BPCS controller. The SIS final element is an independent isolation valve with a three way solenoid as an interface. The BPCS controller is connected to a modulating control valve.

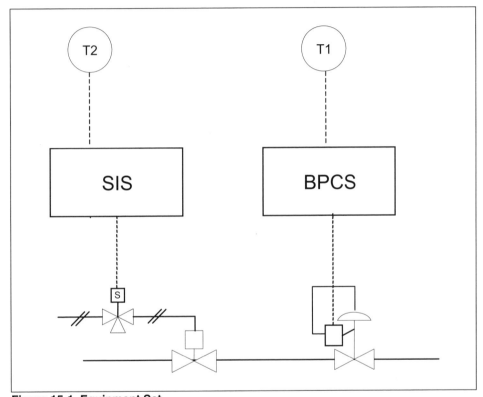

Figure 15-1. Equipment Set

There are many ways of utilizing this equipment to achieve optimal results. For analysis, the data in Tables 15-1, 15-2, 15-3, and 15-4 are used. It is also assumed that the SIS logic solver is programmed to detect out-of-range analog signals and send a diagnostic alarm.

Combined BPCS/SIS Designs

Table 15-1. Sensor Data (Ref. 3)

Component	Failure Rate [1/h]	MTTF (years)	% Safe Failures	Safe Coverage Factor (%)	Dangerous Coverage Factor (%)
Generic pressure transmitter in clean service (high trip)	1.50E-06	76	0	0	60
Generic isolator	5.00E-06	22.83	80	0	0

Table 15-2. Logic Solver Data – Generic SIL3 PLC (Ref. 3)

Component	SD Failure Rate (1/h)	SU Failure Rate (1/h)	DD Failure Rate (1/h)	DU Failure Rate (1/h)
Main Processor	7.43E-06	7.50E-08	2.38E-06	1.25E-07
Power Supply	2.25E-06		2.50E-07	
Analog In Module	9.90E-07	1.00E-08	9.00E-07	1.00E-07
Analog In Channel	4.80E-08	3.00E-09	4.80E-08	3.00E-09
Digital In Module	5.70E-07	3.00E-08	3.80E-07	2.00E-08
Digital In Channel	1.24E-07	7.00E-09	6.70E-08	4.00E-09
Analog Out Module	1.43E-06	7.50E-08	4.75E-07	2.50E-08
Analog Out Channel			9.50E-08	5.00E-09
Digital Out Low Module	7.60E-07	4.00E-08	1.90E-07	1.00E-08
Digital Out Low Channel	1.39E-07	1.00E-09	5.70E-08	3.00E-09

Table 15-3. Final Element Data (Ref. 3)

Component	Failure Rate [1/h]	MTTF (years)	% Safe Failures	Safe Coverage Factor (%)	Dangerous Coverage Factor (%)
Generic 3-way solenoid	6.00E-06	19	60	0	0
Generic air operated ball valve	3.00E-06	38	55	0	0
Generic Control Valve	2.50E-06	45	40	0	0

Table 15-4. Model Parameters

Plant startup time after a trip	24 Hours
Maximum time between plant turnarounds	5 Years
Beta Factor (for valve and actuator)	8%
Beta Factor (for logic solver)	2%
Beta Factor (for transmitters)	5%
MTTR	8 Hours
Proof Test Interval	1 Year
Proof Test Coverage	90%

Approach 1 – Separate BPCS and SIS

Most designers would choose a completely separate BPCS and SIS as shown in Figure 15-1. In this design option, the BPCS could be used as an independent layer of protection, assuming that a shutdown function was included in the control logic. Although the BPCS can only provide a risk reduction factor of 10, it effectively lowers the SIL requirement for the SIF. The SIF itself can be analyzed for risk reduction. For comparison, a combined risk reduction factor can be calculated. The results are shown in Table 15-5.

Table 15-5. Separate SIS Results

	PFDavg	MTTFS	RRF
Sensor subsystem	3.67E-03	764	
Logic solver subsystem	3.25E-05	201	
Final element subsystem	1.93E-02	22	
Total SIF	2.29E-02	19.5	44
Equipment Set			440

The results of the SIF analysis show a PFDavg that equals 0.0229. This provides a risk reduction factor of 44. Given the BPCS layer of protection, the total risk reduction factor of the equipment set is 440.

Approach 2 – Using BPCS transmitter as a diagnostic in the SIS

By adding a loop isolator, the BPCS transmitter signal can be sent to the SIS as shown in Figure 15-2. It may be used as a second shutdown signal or it may be used as a comparison diagnostic in the SIS. In the next design option, the signal will be used as a diagnostic only. It will not be part of the safety functionality. However, since common equipment is used the BPCS cannot be considered an independent layer of protection.

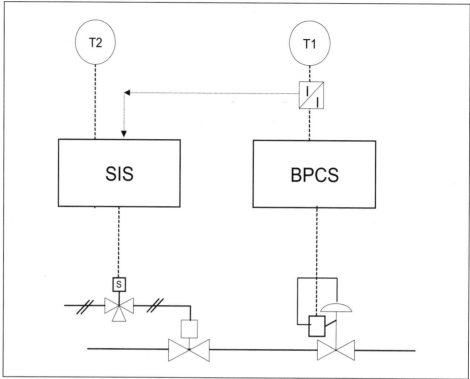

Figure 15-2. Using Second Transmitter as a Comparison Diagnostic

The results of this analysis show improvement in the sensor subsystem. The results are shown in Table 15-6.

Table 15-6. SIF with Sensor Comparison Results

	PFDavg	MTTFS	RRF
Sensor subsystem	1.95E-04	761	
Logic solver subsystem	3.32E-05	201	
Final element subsystem	1.93E-02	22	
Total SIF	1.95E-02	19.5	51
Equipment Set			51

The results of the SIF analysis show a PFDavg that equals 0.00195. This provides a risk reduction factor of 51.

Approach 3 – Using BPCS transmitter and valve as a 1oo2 trip architecture
In the next design option, the transmitter signal will be used as a second shutdown input with 1oo2 voting and the control valve will be used as a second shutdown mechanism. This is shown in Figure 15-3. common equipment is used, the BPCS cannot be considered an independent layer of protection.

It is assumed that the control valve will have sufficient tight shutoff performance to protect against the hazard by itself. If this is not the case, then this design cannot be used.

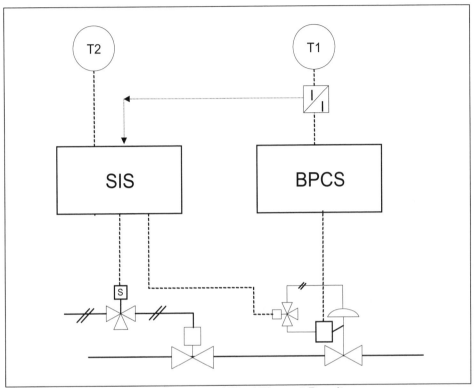

Figure 15-3. 1oo2 Voting for Sensor and Final Element Results

The results of this analysis show improvement in the SIF, primarily in the final element. The results are shown in Table 15-7.

Table 15-7. 1oo2 Voting SIF Results

	PFDavg	MTTFS	RRF
Sensor subsystem	2.45E-04	313	
Logic solver subsystem	3.41E-05	196	
Final element subsystem	2.32E-03	12	
Total SIF	2.60E-03	11	384
Equipment Set			384

The results of the SIF analysis show a PFDavg that equals 0.0026. This provides a risk reduction factor of 384. This is a significant improvement due mainly to the final element. The overall results however are similar to the original separate design. Given that this approach is not significantly better than the separate system, this more complex approach is not the preferred method. In most cases, the separation of control and safety is the correct and strongly preferred solution.

Detailed Analysis of Combination BPCS and SIS Systems

It is certainly possible to do more detailed analysis of combined BPCS and SIS systems. It is possible to identify exactly which components are independent and which components are in common. A model showing the PFDavg and MTTFS of the combined system can be created. Consider the final element assembly from the BPCS as shown in Figure 15-4.

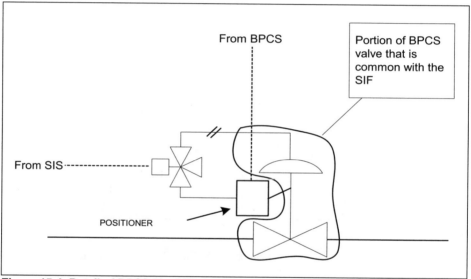

Figure 15-4. Detail of BPCS Final Element

In this situation, the solenoid is an independent part of the SIF, as it is not required or used for the control. The positioner is used only for the control and is not part of the SIF. Only the actuator and the valve are common to both BPCS and SIF. A detailed model showing these exact relationships can be created for those who need more accuracy.

Exercises

15-1. Two different types of solenoid valves are used to block fuel flow to a burner in a SIS. The valves are piped in series. Both valves should close when a dangerous condition is detected. Both valves have one failure mode, fail-danger, with a failure rate of 0.0008 failures per year. Both valves are tested once every year. Based on the differences between the valves, a common cause beta factor of 0.01 is assigned. What is the PFD of the valve subsystem including common cause?

   a. 0.0008
   b. 0.001
   c. 0.000072
   d. 0.0000017
   e. 0.999

15-2. A designer wants to use two pressure transmitters in a 1oo2 architecture to achieve a SIL 3 safety function. The individual pressure transmitters have been certified per IEC 61508 to be fit for use in a SIL 2 safety application. The PFDavg calculation for the 1oo2 safety loop shows that the PFDavg is within the SIL 3 range. Explain why this parallel configuration may not necessarily meet the SIL 3 objectives according to IEC 61508 / 61511 standards.

15-3. A "smart" transmitter has a failure rate of 0.06 failures/year. The safe failures ratio is 66%, and the diagnostic coverage of dangerous failures is 60%. The diagnostic coverage for safe failures is 70%. Assuming all detected dangerous failures will immediately be converted to a safe process shutdown, what is the maximum probability of failure on demand if the transmitter is tested twice per year?

   a. 0.0041
   b. 0.05
   c. 0.66
   d. 0.7
   e. 0.0003

15-4. Repair time for a certain failure is estimated to average 1 hour. What is the maximum time increment for a discrete time Markov model?

   a. 1 hour
   b. 2 hours
   c. 0.5 hours
   d. 0.1 hours

15-5. A failure rate is given as 0.15 failures per year. What is the probability of failure for a Markov model with a one hour time increment?

   a. 0.0002
   b. 0.2
   c. 0.028
   d. 0.000017

REFERENCES AND BIBLIOGRAPHY

1. ANSI/ISA-84.00.01-2004, *Functional Safety: Safety Instrumented Systems for the Process Industry Sector – Parts 1, 2, and 3 (IEC 61511 Mod)*. ISA, 2004.

2. IEC 61508, *Functional Safety of electrical / electronic / programmable electronic safety-related systems*, IEC, 2000.

3. *Safety Equipment Reliability Handbook.* exida, 2003. (available from ISA)

Appendix A
Statistics

Random Variables

Many processes have outcomes that cannot be predicted given our current level of knowledge about the process. In such situations statistical analysis can be used to gain knowledge about a process from a set of data. Data is gathered by recording a specific random variable. Statistical analysis provides specific information about that random variable. In reliability engineering the primary random variable is "time to failure,' the successful operating time interval until a failure occurs.

Statistical analysis is quite useful because data, when gathered, is often hard to understand. Consider a set of data shown in Table A-1. This set of data is a record of failure times for thirty systems. Assume that all thirty systems are installed, commissioned and operating successfully. The units are checked every hour and the total number of hours of successful operating time is incremented. When a particular system fails, the successful operating time is no longer incremented. For example in this data set, system one failed after 96 hours. System two failed after 3091 hours. System thirty failed after a successful operating time interval of 409 hours. A set of data exists, but often the useful information hides inside the data.

One can study the data and gain insight regarding when a system might fail. It is notable that system 12 failed after only 33 hours. This system failed first and had the shortest successful operating time interval. Several systems had much longer successful operating times. System 17 had the longest successful operating time. It ran for 13990 hours. It is hard to really gain an in depth understand from just looking at the raw data. Fortunately, one can apply some statistical analysis to gain further insight.

Table A-1. Time To Failure Data Set

System	Hours	System	Hours
1	96	16	1282
2	3091	17	13990
3	4862	18	12751
4	13853	19	2106
5	8339	20	5431
6	614	21	2740
7	1815	22	11460
8	10305	23	6056
9	7499	24	3471
10	1540	25	2414
11	831	26	4348
12	33	27	3886
13	240	28	9270
14	196	29	13351
15	1045	30	409

Statistical Analysis

Often, statistical analysis is done by grouping data into "bins." For example, the failure data may track the number of units failed in one thousand hour increments (Table A-2). In many cases the data is gathered this way as operation is checked periodically. When successful operation is only checked every one thousand hours, the data set would naturally look like Table A-2.

An examination of the data in this format shows additional information. One can observe that more systems fail in the first block of time than any other. It is also clear that the quantity of systems that fail in each block decreases with increasing time. After that the quantity of systems that fail in each time block remains rather constant until the last block. This form of data is often presented graphically in a form called a histogram, Figure A-1. Many consider the graphical format of the histogram to be a very effective to quickly represent the information.

Table A-2. Failure Data Grouped into One Thousand Hour Increments

Hours	Units	Cum.
0-1000	7	7
1001-2000	4	11
2001-3000	3	14
3001-4000	3	17
4001-5000	2	19
5001-6000	1	20
6001-7000	1	21
7001-8000	1	22
8001-9000	1	23
9001-10000	1	24
10001-11000	1	25
11001-12000	1	26
12001-13000	1	27
13001-14000	3	30

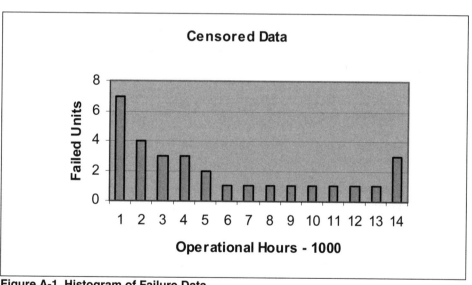

Figure A-1. Histogram of Failure Data

An assignment of probability can be made based on this data set. The histogram in Figure A-1 is "normalized" by dividing each data quantity by the total. In this way a histogram is converted into a "probability density function (pdf)." A pdf is a plot of probability versus the statistical variable. In this example, the statistical variable is operational failure time.

Figure A-2 shows the pdf for this set of data. The probability of a failure in the first time period (0-1000) hours is 0.233.

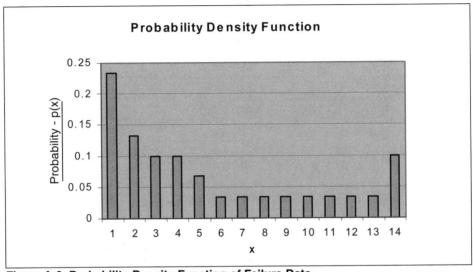

Figure A-2. Probability Density Function of Failure Data

Mean – Median – Standard Deviation

Statistical analysts can provide numbers that characterize the data set. Two of the most common numbers provided are called the "mean" and the "median." The mean is an arithmetic average of the data. It provides insight into the data set.

> **EXAMPLE A-1**
>
> **Problem:** A set of time to failure data is given in Table A-1. What is the mean of this data?
>
> **Solution:** The numbers are summed and divided by the total quantity of systems (30).
>
> $$\text{Mean} = 147{,}324 / 30 = 4{,}910.8$$
>
> The median is the "middle" value. For an even quantity of data, the median is average of the two middle numbers.

The difference between the mean and the median indicates the symmetry of the probability density function (pdf). For a symmetrical distribution like a normal distribution, the numbers would be similar. For this example the difference in the two numbers indicates a non-symmetric pdf. This can be seen in Figure A-2.

> **EXAMPLE A-2**
>
> **Problem:** A set of time to failure data is given in Table A-1. What is the median of this data?
>
> **Solution:** Median = (3091+3471)/2 = 3281 Hours

Another common number calculated for data sets is called the standard deviation. This number indicates the number range of the pdf. In failure data analysis this number is not commonly used as the failure rate pdf is characterized by other numbers.

Statistical Significance – Applicability

Two things that should be remembered about all statistics, and especially failure statistics, are statistical significance and applicability. The more data we have, the more certain we are that the statistics have meaning. That measure has been fully developed and is called statistical significance. With little data, one cannot be sure about the accuracy of the result.

One must also be concerned about applicability of the statistics. This is especially relevant regarding failure data. If the statistical data set was obtained under conditions that are completely different than current conditions, one must ask if the data is relevant.

Many statisticians will take a "Bayesian" approach to the problem estimating the probability that a data set is valid. Given a number of data sets, effectively an average of the data sets is obtained.

> **EXAMPLE A-3**
>
> **Problem:** A set of time to failure data has a mean of 5521 hours. A second set of time to failure data has a mean of 4911 hours. A third set of time to failure data has a mean of 12,340 hours. It is estimated that the first data set has a 50% chance of being correct. It is estimated that the second set of data has a 40% chance of being correct. It is estimated that the third set of data has a 10% chance of being correct. What is most likely value for the mean?
>
> **Solution:**
>
> Expected value of Mean = $5521 \times 0.5 + 4911 \times 0.4 + 12340 \times 0.1 = 5959$

This tells us that one must record not only the data under study but all suspected relevant conditions.

Appendix B
Probability

Probability Assignment

Probability is a quantitative method of expressing chances. A probability is assigned a number between zero and one, inclusive. A probability assignment of zero means that the event is never expected. A probability assignment of one means that the event is always expected.

Probabilities are often assigned based on historical "frequency of occurrence." An experiment is repeated many times, say N. A quantity is tabulated for each possible outcome of the experiment. For any particular outcome, the probability is determined by dividing the number of occurrences, n, by the number of trials.

$$P(E) = \frac{n}{N} \tag{B-1}$$

The values become more certain as the number of trials is increased. A definition of probability based on this concept is stated in Equation B-2:

$$P(E) = \lim_{N \to \infty} \frac{n}{N} \tag{B-2}$$

Venn Diagrams

A convenient way to depict the outcomes of an experiment is through the use of the Venn diagram. These diagrams were created by John Venn (1834-1923), an English mathematician and cleric. They provide visual representation of data sets, including experimental outcomes. The diagrams are drawn by using the area of a rectangle to represent all possible outcomes; this area is known as the "sample space." Any particular outcome is shown by using a portion of the area within the rectangle.

For the toss of a fair pair of dice, the possible outcomes are shown in the Venn diagram of Figure B-1. The outcomes do not occupy the same area on the diagram. The probabilities of some outcomes are more likely than others; these occupy more area. For example, the area occupied by an outcome of "2" is 1/36 of the total. The area occupied by the outcome "7" is 6/36 of the total. Again, the area occupied by each outcome is proportional to its probability.

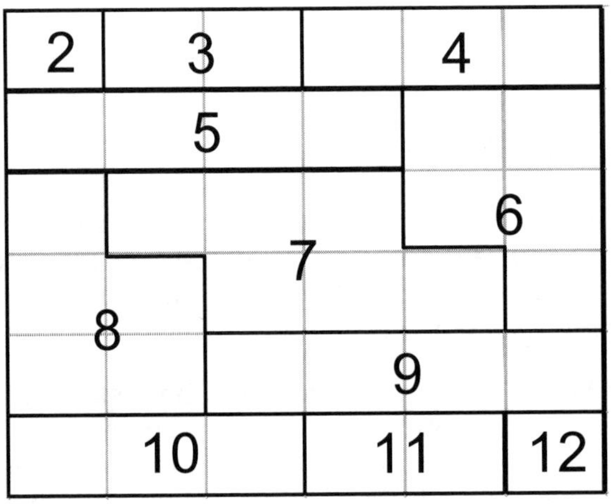

Figure B-1. Venn Diagram - Dice Toss

A Venn diagram is often used to identify the attributes of possible outcomes. Outcomes are grouped into sets that are based on some characteristic or combination of characteristics. The graphical nature of the Venn diagram is especially useful in showing these combinations of sets: unions, intersections, and complementary events.

A *Union* of some number of sets (A, B, C) is defined as any event in either set A or set B or set C. This is represented in a Venn diagram as shown in Figure B-2. Vertical lines extend through three circles, the A circle, the B circle, and the C circle.

An *Intersection* of sets (A, B, C) is defined as any event in sets A and B and C. This is represented in a Venn diagram as shown in Figure B-3. Only one small area is in all three circles -- that area is marked with lines.

Complementary sets are easily shown on Venn diagrams. Since the diagram represents the entire sample space, all area not enclosed within an event is

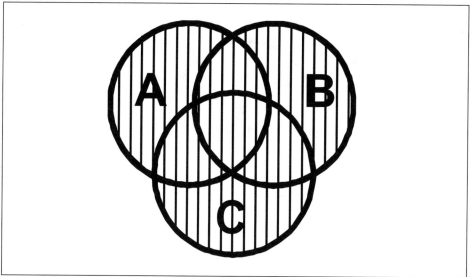

Figure B-2. Union of Sets A, B, and C

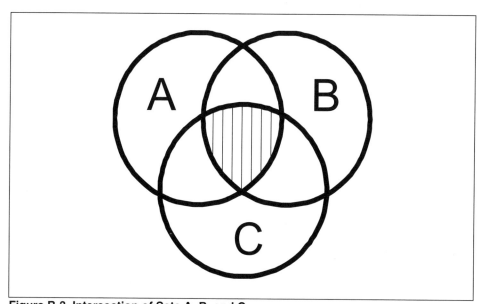

Figure B-3. Intersection of Sets A, B, and C

the complement of the event. In Figure B-4, a circle represents the set A. Its complement is set B, represented by the remainder of the diagram.

Mutually exclusive sets are defined as sets that cannot happen at the same time. Mutually exclusive event sets are easily recognized on a Venn diagram. In Figure B-5, the event sets A and B are shown. There is no common space within the A circle and the B circle. There is no intersection between A and B. They cannot happen at the same time and are, therefore, mutually exclusive.

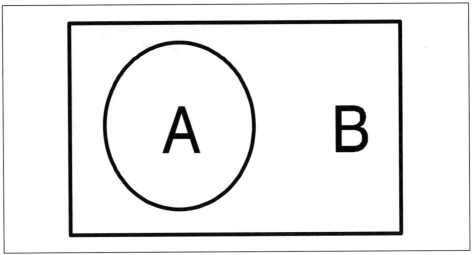

Figure B-4. Complementary Events

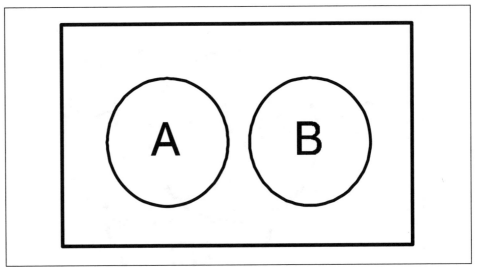

Figure B-5. Mutually Exclusive Events

Combining Probabilities

Certain rules help to calculate the probability of combinations of events. Combinations of events are common in the field of reliability evaluation. Often system failures occur only when certain combinations of events happen during certain times.

Independent Events

If the occurrence of an event from set A does not affect the probability of events from set B, then sets A and B are defined to be "independent"; for example, the outcome of one coin toss does not affect the next toss. The outcome of one coin toss does not give us any information about the outcome of a subsequent independent coin toss. When two event sets are independent, the probability of getting an event from set A and set B (the intersection) is given by the formula:

$$P(A \cap B) = P(A) \times P(B) \qquad (B\text{-}3)$$

Independent events are different than mutually exclusive events. Consider two events, A and B that are mutually exclusive. Knowing that A has occurred tells us that B cannot occur. If events A and B are independent, knowing that A occurred tells us nothing about B. Two events, A and B, cannot be both mutually exclusive and independent.

EXAMPLE B-1

Problem: Two fair coins are flipped into the air. What is the probability that both coins will land with heads showing?

Solution: Each coin toss has only two possible outcomes: heads or tails. Each outcome has a probability of one half. The coin tosses are independent. Therefore,

$$P(\text{two heads}) = P(\text{head 1}) \times P(\text{head 2}) = 1/2 \times 1/2 = 1/4$$

EXAMPLE B-2

Problem: A pair of fair (well balanced) dice is rolled. What is the probability of getting "snake eyes" -- one dot on each die?

Solution: The outcome of one die does not affect the outcome of the other die. Therefore, the events are independent. The probability of getting one dot can be obtained by noting that there are six sides on the die and that each side is equally likely. The probability of getting one dot is one sixth (1/6). The probability of getting "snake eyes" is represented as:

$$P(1,1) = 1/6 \times 1/6 = 1/36$$

Check the area occupied by the "2" result on Figure B-1. Is that area equal to one thirty-sixth?

Probability Summation

If the probability of getting a result from set A equals 0.2 and the probability of getting a result from set B equals 0.3, what is the probability of getting a result from either set A or set B?

> **EXAMPLE B-3**
>
> **Problem:** A controller fails only if the input power fails and the controller battery fails. Assume that these factors are independent. For a time interval of five years, the probability of input power failure is 0.0001 and the probability of battery failure is 0.01. What is the probability of controller failure during the time interval of five years?
>
> **Solution:** Since input power and battery failure are independent, Equation B-3 gives the probability of both events:
>
> $$P(\text{Controller Failure}) = 0.0001 \times 0.01 = 0.000001$$

It would be natural to assume that the answer is 0.5, the sum of the above probabilities, but that answer is not always correct. Look at the Venn diagram in Figure B-6. If the area of set A (6/36) is added to the area of set B (6/36), the answer (12/36) is too large. (The answer should be 11/36.) Since there is an intersection between sets A and B, the area of the intersection has been counted twice. When summing probabilities, the intersections must be subtracted. Thus, the probability of the union of event sets A and B is given by:

$$P(A \cup B) = P(A) + P(B) - P(A \cap B) \tag{B-4}$$

If set A and set B are mutually exclusive so there is no intersection, then the following can be stated.

$$P(A \cup B) = P(A) + P(B) \tag{B-5}$$

> **EXAMPLE B-4**
>
> **Problem:** A pair of fair dice is rolled. What is the probability of getting a sum of seven?
>
> **Solution:** A sum of seven dots on the dice can be obtained in a number of different ways; these are described by the sets {1,6}, {2,5}, {3,4}, {4,3}, {5,2}, and {6,1}. Each specific combination has a probability of 1/36. The combinations are mutually exclusive; therefore, Equation B-5 can be used.
>
> $$P(\text{seven dots}) = 1/36 + 1/36 + 1/36 + 1/36 + 1/36 + 1/36 = 1/6$$

Conditional Probability

Often it is required to calculate the probability of some event under specific circumstances. The probability of event A, given that event B has occurred, may need to be calculated. Such a probability is called a

EXAMPLE B-5

Problem: A pair of fair dice is rolled. What is the probability of getting an even number on both dice?

Solution: On each die there are six numbers. Three of the numbers are odd (1, 3, 5) and three of the numbers are even (2, 4, 6). All numbers are mutually exclusive. Equation B-5 gives the probability of getting an even number on one die.

P(even) = P(2,4,6) = P(2) + P(4) + P(6) = 1/6 + 1/6 + 1/6 = 1/2

The outcome of one die is independent of the other die. Therefore,

P(even, even) = P(Set A even) × P(Set B even) = 1/2 × 1/2 = 1/4

EXAMPLE B-6

Problem: A pair of fair dice is rolled. What is the probability of getting two dots on either or both dice?

Solution: The probability of getting two dots on die A or B equals 1/6. The probability of getting two dots on both dice though is 1/36. Because these events of independent, they are not mutually exclusive. Therefore, we can use Equation B-4.

P(A OR B) = P(A) + P(B) − P(A × B) = 1/6 + 1/6 − 1/36 = 11/36

This is evident in Figure B-6, a Venn diagram of the problem.

"conditional probability." The situation can be envisioned by examining the Venn diagram in Figure B-7.

Normally, the probability of event A would be given by the area of circle A divided by the total area. Conditional probability is different. For example, consider the situation when Event B has occurred. This means that only the state space within the area of circle B needs to be examined. This is a substantially reduced area! The desired probability is the area of circle A within circle B, divided by the area of circle B, expressed by:

$$P(A|B) = \frac{P(A \cap B)}{P(B)} \tag{B-6}$$

This reads: the probability of A, given B, is equal to the probability of the intersection of A and B divided by the probability of B. The area of circle A within circle B represents the probability of the intersection of A and B. The area of circle B equals the probability of B.

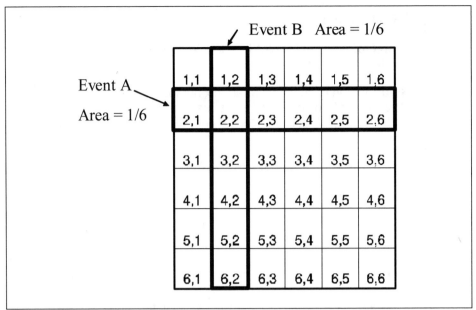

Figure B-6. Venn Diagram of Example B-6

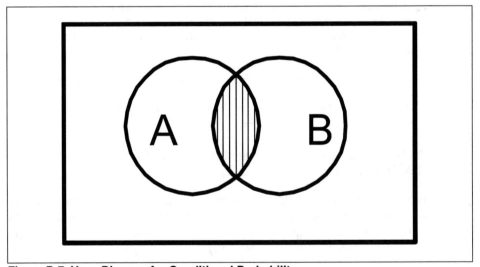

Figure B-7. Venn Diagram for Conditional Probability

The formula for conditional probability can be rearranged as:

$$P(A \cap B) = P(A \mid B) \times P(B) \qquad (B-7)$$

This states that the intersection of events A and B can be obtained by multiplying the probability of A, given B, times the probability of B. When the statistics are kept in a conditional format, this equation can be useful.

EXAMPLE B-7

Problem: A pair of fair dice is rolled. What is the probability of getting a two on both dice given that one die has a two?

Solution: The probability of {2,2}, given that one die has a two, is given by Equation B-6:

$$P(2,2) = 1/36 \,/\, 1/6 = 1/6$$

In this case, the answer is intuitive since the outcome of each die is independent.

EXAMPLE B-8

Problem: A pair of fair dice is rolled. What is the probability of getting a sum of seven, given that exactly one die shows a two?

Solution: There are only two ways to get a sum of seven, given that one die has a two. Those two combinations are {2,5} and {5,2}. There are 10 combinations that show a two on exactly one die. These sets are {2,1}, {2,3}, {2,4}, {2,5}, {2,6}, {1,2}. {3,2}, {4,2}, {5,2}, and {6,2}. Using Equation B-6:

$$P(A \mid B) = \frac{P(A \cap B)}{P(B)}$$

$$= \frac{\frac{2}{36}}{\frac{10}{36}}$$

$$= \frac{2}{10}$$

This is shown graphically in Figure B-8.

2,1	2,3	2,4	2,5	2,6
1,2	3,2	4,2	5,2	6,2

Figure B-8. Probability Diagram of Example B-8

Bayes' Rule

Consider an event A. The state space in which it exists is divided into two mutually exclusive sections, B and B' (Figure B-9). Event A can be written as:

$$A = (A \cap B) \cup (A \cap B') \tag{B-8}$$

Since AB and AB' are mutually exclusive,

$$P(A) = P(A \cap B) + P(A \cap B') \tag{B-9}$$

Substituting Equation B-7 into B-9,

$$P(A) = P(A|B) \times P(B) + P(A|B') \times P(B') \tag{B-10}$$

This states that the probability of event A equals the conditional probability of A, given that B has occurred, plus the conditional probability of A, given that B has not occurred. This is known as Bayes' rule. It is used in many aspects of reliability engineering.

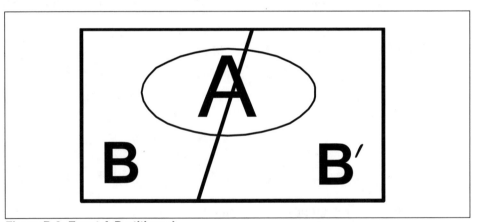

Figure B-9. Event A Partitioned

EXAMPLE B-9

Problem: The workday is divided into three mutually exclusive time periods: day shift, evening shift, and night shift. Day shift lasts ten hours. Evening shift is eight hours. Night shift is six hours. Logs show that in the last year (8760 hours) one failure occurred during the day shift (one failure in 3650 hours), two failures occurred during the evening shift (two failures in 2920 hours), and seven failures occurred during the night shift (seven failures in 2190 hours). What is the overall probability of failure?

Solution: Define event A as failure. Define event B1 as the day shift, B2 as the evening shift, and B3 as the night shift. The probability of failure given, event B1 (day shift) is calculated knowing that one failure occurred in 3650 hours (one third of the hours in one year). A variation of Equation B-10 can be used where P(B1) is day shift probability, P(B2) is the evening shift probability, and P(B3) is the night shift probability.

$$P(fail) = P(fail \mid B1) \times P(B1) + P(fail \mid B2) \times P(B2) + P(fail \mid B3) \times P(B3)$$

The probabilities of failure for each shift are calculated by dividing the number of failures during each shift by the numbers of hours in each shift. Substituting the numbers:

$$P(fail) = (1/3650 \times 10/24) + (2/2920 \times 8/24) + (7/2190 \times 6/24)$$

$$= 0.000114 + 0.000226 + 0.000799 = 0.001139$$

EXAMPLE B-10

Problem: A company manufactures controllers at two locations. Sixty percent are manufactured in plant X. Forty percent are manufactured in plant Y. Controllers manufactured in plant X have a 0.00016 probability of failure in a one year period. Controllers manufactured in plant Y have a 0.00022 probability of failure in one year. A purchased controller can come randomly from either source. What is the probability of a controller failure?

Solution: Define controller failure as event A. Define event B1 as plant X manufacture. Define event B as plant Y manufacture. Using Equation B-10, substitute the values to obtain:

$$P(fail) = (0.00016 \times 0.6) + (0.00022 \times 0.4) = 0.000096 + 0.000088 = 0.000184$$

Appendix C
Fault Trees

Solution Techniques

Fault Trees and Reliability Block Diagrams are both methods of showing probability combinations. There have been a number of solution techniques developed to solve probability combinations. These include Cut Sets, Tie sets, Event Space, Decomposition Method, Gate Solution Method, and many others. In this appendix three examples will be shown – the Event Space method and the Cut Set method, and the Gate Solution Method. Details and full development of the methods can be found in (Ref. 1) Chapter 5.

This chapter will also discuss and explain the potential problems of PFavg calculation with any probability combination method like fault trees. Probability averages must be taken after logic is applied, or else significant errors may result.

Event Space Method

One technique that provides a comprehensive overview of a probability combination model is the event space method. The authors have used this technique especially when a detailed knowledge of probability combinations is needed to optimize a design. The method is called systematic and comprehensive.

Consider the example of a 2oo3 redundant power supply system as shown in Figure C-1. The system consists of three diverse power supplies each capable of providing 100 watts of power. The total load on the system is 175 watts. Therefore, the system can be successful in delivering 175 watts if only two of the three power subsystems are available. The fault tree for this system is shown in Figure C-2 and a reliability block diagram is shown in Figure C-3.

Appendix C: Fault Trees

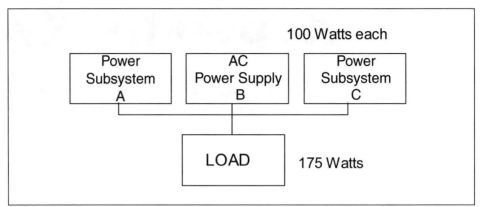

Figure C-1. Two out of Three Power System

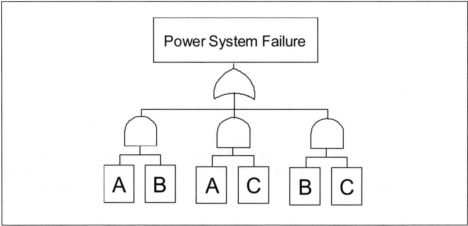

Figure C-2. Fault Tree for 2oo3 Power System

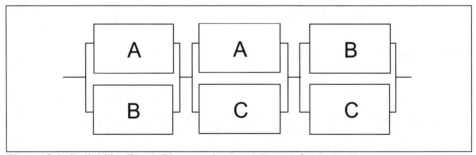

Figure C-3. Reliability Block Diagram for 2oo3 Power System

To solve the fault tree or the reliability block diagram via the event space method one creates a list of all combinations of successful and failed components. For each line in the list, the probability is listed and the system operation (success or failure) is listed. By convention, this is done by quantity of failed units.

> **EXAMPLE C-1**
>
> **Problem:** A 2oo3 power system has three subsystems. The probability of failure during the next year for subsystem A is 0.03. The probability of failure during the next year for subsystem B is 0.008. The probability of failure during the next year for subsystem C is 0.02. What is the probability of failure for the system?
>
> **Solution:** Using the event space method a table is created showing all combinations of successful and failed components starting with all components successful. This is shown in Table C-1.

Table C-1. Event Space Analysis for 2oo3 Power System

			Subsystem Probability			State Probability	System Fail	
			A	B	C			
Group 0 - no failures								
A OK	B OK	C OK	0.97	0.992	0.98	0.9429952	0	0
Group 1 - one failure								
A Fail	B OK	C OK	0.03	0.992	0.98	0.0291648	0	0
A OK	B Fail	C OK	0.97	0.008	0.98	0.0076048	0	0
A OK	B OK	C Fail	0.97	0.992	0.02	0.0192448	0	0
Group 2 - two failures								
A OK	B Fail	C Fail	0.97	0.008	0.02	0.0001552	1	0.000155
A Fail	B OK	C Fail	0.03	0.992	0.02	0.0005952	1	0.000595
A Fail	B Fail	C OK	0.03	0.008	0.98	0.0002352	1	0.000235
Group 3 - three failures								
A Fail	B Fail	C Fail	0.03	0.008	0.02	0.0000048	1	4.8E-06
							Total	0.00099

Group 0 is the situation where none of the subsystems have failed. The probability of subsystem A being successful during a one year interval is 0.97 (1 − 0.03). The probability of subsystem B being successful during a one-year interval is 0.992 (1 − 0.008). The probability of subsystem C being successful during a one-year interval is 0.98 (1 − 0.02). The probability that the system is in this state is the AND of all three subsystem conditions. In this case it is equal to $0.97 \times 0.992 \times 0.98 = 0.9429952$. This calculation is continued for each row.

The "System Fail" column indicates which combinations of subsystem failure represent system failure. Those combinations are added to obtain system failure probability. For this example, the probability of system failure for a one-year interval equals 0.00099.

Cut Set Method

The "Cut Set" method was named from the reliability block diagram. Notice in Figure C-3 that if one "cuts" across a set of blocks in parallel that the system fails. The reliability block diagram is draw in the "cut set" style to show those cut sets.

A cut set is defined as "a set of components that when failed, will fail the system." A minimal cut set is defined as "a set of components that when failed, will fail the system such that if any one of them is successful, the system is successful." The objective of the analysis is to determine the minimal cut set. From the minimal cut set, the probability of system failure is determined.

EXAMPLE C-2

Problem: A 2oo3 power system has three subsystems, A, B, and C. The set of failure events ABC will fail the system. Is this a cut set? Is this a minimal cut set?

Solution: If subsystem A, B and C fail, the system will fail. This set of failure events meets the definition of a cut set. However, that set does not meet the definition of a minimal cut set as one of those subsystems can be restored and the system is still failed.

EXAMPLE C-3

Problem: Using the cut set method, solve for probability of system failure for the 2oo3 power system using the failure probabilities from EXAMPLE C-1.

Solution: Probabilities can be combined using the AND function for each cut set. However, the union of the cut sets cannot be simply added as cut set probabilities are not mutually exclusive and the same failure event will likely appear in more than one cut set. Unions (OR gates) still apply.

The cut set probabilities are given by:

$$C1 = P(A\ Fail) \times P(B\ Fail)$$

$$C2 = P(A\ Fail) \times P(C\ Fail)$$

$$C3 = P(B\ Fail) \times P(C\ Fail)$$

The union is:

$$P(System\ Fail)$$

$$= P(C1\ U\ C2\ U\ C3)$$

$$= P(C1) + P(C2) + P(C3) - P(C1 \times C2) - P(C1 \times C3) - P(C2 \times C3) + P(C1 \times C2 \times C3)$$

Where

$$P(C1 \times C2) = P(A\ Fail) \times P(B\ Fail) \times P(C\ Fail)$$

Example C-3 continued

Not P(A Fail) × P(A Fail) × P(B Fail) × P(C Fail) as one would get if merely multiplying probabilities. This is because the probability of getting A AND A is A.

$$P(C1 \times C3) = P(A\ Fail) \times P(B\ Fail) \times P(C\ Fail)$$

$$P(C2 \times C2) = P(A\ Fail) \times P(B\ Fail) \times P(C\ Fail)$$

And

$$P(C1 \times C2 \times C3) = P(A\ Fail) \times P(B\ Fail) \times P(C\ Fail)$$

The result is:

$$P(System\ Fail) = P(A\ Fail) \times P(B\ Fail) + P(A\ Fail) \times P(C\ Fail) + P(B\ Fail) \times P(C\ Fail) - 2 \times P(A\ Fail) \times P(B\ Fail) \times P(C\ Fail)$$

Substituting the probabilities:

$$P(System\ Fail) =$$

$$(0.03 \times 0.008) + (0.03 \times 0.02) + (0.008 \times 0.02) - 2(0.03 \times 0.008 \times 0.02) =$$

$$0.00024 + 0.0006 + 0.00016 - 2 \times 0.0000048 =$$

$$0.00099$$

EXAMPLE C-4

Problem: A fire sensor subsystem uses four sensors. If any two of the four sensors indicate a fire then an alarm will be sounded. There is a probability that a sensor will fail to indicate a fire for a one year interval of 0.01. What is the probability that the sensor subsystem will fail to indicate a fire?

Solution: A fault tree can be drawn to show the probability combinations for failure to indicate a fire. A fault tree is a good tool to use for the problem compared to a reliability block diagram as this problem focuses on one failure mode, failure to indicate a fire. (Another failure mode might be false indication of a fire.) The fault tree is shown in Figure C-4.

Using the event space method as shown in Figure C-5, the entire picture is laid out for easy checking.

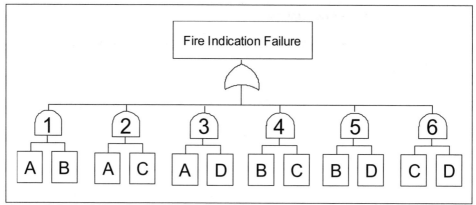

Figure C-4. Fault Tree for Fire Indication Failure

				Subsystem Probability				State	System Fail	
Group 0 - no failures				A	B	C	D	Probability		
A OK	B OK	C OK	D OK	0.99	0.99	0.99	0.99	0.960596	0	0
Group 1 - one failure										
A Fail	B OK	C OK	D OK	0.01	0.99	0.99	0.99	0.009703	0	0
A OK	B Fail	C OK	D OK	0.99	0.01	0.99	0.99	0.009703	0	0
A OK	B OK	C Fail	D OK	0.99	0.99	0.01	0.99	0.009703	0	0
A OK	B OK	C OK	D Fail	0.99	0.99	0.99	0.01	0.009703	0	0
Group 2 - two failures										
A Fail	B Fail	C OK	D OK	0.01	0.01	0.99	0.99	9.8E-05	1	0.00009801
A Fail	B OK	C Fail	D OK	0.01	0.99	0.01	0.99	9.8E-05	1	0.00009801
A Fail	B OK	C OK	D Fail	0.01	0.99	0.99	0.01	9.8E-05	1	0.00009801
A OK	B Fail	C Fail	D OK	0.99	0.01	0.01	0.99	9.8E-05	1	0.00009801
A OK	B Fail	C OK	D Fail	0.99	0.01	0.99	0.01	9.8E-05	1	0.00009801
A OK	B OK	C Fail	D Fail	0.99	0.99	0.01	0.01	9.8E-05	1	0.00009801
Group 3 - three failures										
A Fail	B Fail	C Fail	D OK	0.01	0.01	0.01	0.99	9.9E-07	1	0.00000099
A Fail	B Fail	C OK	D Fail	0.01	0.01	0.99	0.01	9.9E-07	1	0.00000099
A Fail	B OK	C Fail	D Fail	0.01	0.99	0.01	0.01	9.9E-07	1	0.00000099
A OK	B Fail	C Fail	D Fail	0.99	0.01	0.01	0.01	9.9E-07	1	0.00000099
Group 4 - four failures										
A Fail	B Fail	C Fail	D Fail	0.01	0.01	0.01	0.01	1E-08	1	0.00000001
								Total		0.00059203

Figure C-5. Event Space Method for Fire Sensor System

The cut set method could also be used to solve the problem. One must be careful, however, to make sure that the unions are properly calculated. The cut set solution is given by:

P (Fire Indication Failure) = AB + AC + AD + BC + BD + CD

Contribution of Individual Gates

Minus the contribution of the fifteen combinations of two gates at a time

− (AB ∪ AC)	which equals −ABC	Combination of gate 1, gate 2
− (AB ∪ AD)	which equals −ABD	Combination of gate 1, gate 3
− (AB ∪ BC)	which equals −ABC	Combination of gate 1, gate 4
− (AB ∪ BD)	which equals −ABD	Combination of gate 1, gate 5
− (AB ∪ CD)	which equals −ABCD	Combination of gate 1, gate 6
− (AC ∪ AD)	which equals −ACD	Combination of gate 2, gate 3
− (AC ∪ BC)	which equals −ABC	Combination of gate 2, gate 4
− (AC ∪ BD)	which equals −ABCD	Combination of gate 2, gate 5
− (AC ∪ CD)	which equals −ACD	Combination of gate 2, gate 6
− (AD ∪ BC)	which equals −ABCD	Combination of gate 3, gate 4
− (AD ∪ BD)	which equals −ABD	Combination of gate 3, gate 5
− (AD ∪ CD)	which equals −ACD	Combination of gate 3, gate 6
− (BC ∪ BD)	which equals −ABD	Combination of gate 4, gate 5
− (BC ∪ CD)	which equals −BCD	Combination of gate 4, gate 6
− (BD ∪ CD)	which equals −BCD	Combination of gate 5, gate 6

Plus the contribution of the twenty combinations of three gates at a time

+ (AB ∪ AC ∪ AD)	which equals ABCD	Contribution of gate 1, gate 2, gate 3
+ (AB ∪ AC ∪ BC)	which equals ABC	Contribution of gate 1, gate 2, gate 4
+ (AB ∪ AC ∪ BD)	which equals ABCD	Contribution of gate 1, gate 2, gate 5
+ (AB ∪ AC ∪ CD)	which equals ABCD	Contribution of gate 1, gate 2, gate 6
+ (AB ∪ AD ∪ BC)	which equals ABCD	Contribution of gate 1, gate 3, gate 4
+ (AB ∪ AD ∪ BD)	which equals ABD	Contribution of gate 1, gate 3, gate 5
+ (AB ∪ AD ∪ CD)	which equals ABCD	Contribution of gate 1, gate 3, gate 6
+ (AB ∪ BC ∪ BD)	which equals ABCD	Contribution of gate 1, gate 4, gate 5
+ (AB ∪ BC ∪ CD)	which equals ABCD	Contribution of gate 1, gate 4, gate 6
+ (AB ∪ BD ∪ CD)	which equals ABCD	Contribution of gate 1, gate 5, gate 6
+ (AC ∪ AD ∪ BC)	which equals ABCD	Contribution of gate 2, gate 3, gate 4
+ (AC ∪ AD ∪ BD)	which equals ABCD	Contribution of gate 2, gate 3, gate 5
+ (AC ∪ AD ∪ CD)	which equals ACD	Contribution of gate 2, gate 3, gate 6
+ (AC ∪ BC ∪ BD)	which equals ABCD	Contribution of gate 2, gate 4, gate 5
+ (AC ∪ BC ∪ CD)	which equals ABCD	Contribution of gate 2, gate 4, gate 6
+ (AC ∪ BD ∪ CD)	which equals ABCD	Contribution of gate 2, gate 5, gate 6
+ (AD ∪ BC ∪ BD)	which equals ABCD	Contribution of gate 3, gate 4, gate 5
+ (AD ∪ BC ∪ CD)	which equals ABCD	Contribution of gate 3, gate 4, gate 6
+ (AD ∪ BD ∪ CD)	which equals ABCD	Contribution of gate 3, gate 5, gate 6
+ (BC ∪ BD ∪ CD)	which equals BCD	Contribution of gate 4, gate 5, gate 6

Minus the contribution of fifteen combinations of four gates at a time

− (AB ∪ AC ∪ AD ∪ BC)	which equals ABCD	Gate 1, 2, 3, 4
− (AB ∪ AC ∪ AD ∪ BD)	which equals ABCD	Gate 1, 2, 3, 5
− (AB ∪ AC ∪ AD ∪ CD)	which equals ABCD	Gate 1, 2, 3, 6
− (AB ∪ AC ∪ BC ∪ BD)	which equals ABCD	Gate 1, 2, 4, 5
− (AB ∪ AC ∪ BC ∪ CD)	which equals ABCD	Gate 1, 2, 4, 6
− (AB ∪ AC ∪ BD ∪ CD)	which equals ABCD	Gate 1, 2, 5, 6
− (AB ∪ AD ∪ BC ∪ BD)	which equals ABCD	Gate 1, 3, 4, 5
− (AB ∪ AD ∪ BC ∪ CD)	which equals ABCD	Gate 1, 3, 4, 6
− (AB ∪ AD ∪ BD ∪ CD)	which equals ABCD	Gate 1, 3, 5, 6
− (AB ∪ BC ∪ BD ∪ CD)	which equals ABCD	Gate 1, 4, 5, 6
− (AC ∪ AD ∪ BC ∪ BD)	which equals ABCD	Gate 2, 3, 4, 5
− (AC ∪ AD ∪ BC ∪ CD)	which equals ABCD	Gate 2, 3, 4, 6
− (AC ∪ AD ∪ BD ∪ CD)	which equals ABCD	Gate 2, 3, 5, 6
− (AC ∪ BC ∪ BD ∪ CD)	which equals ABCD	Gate 2, 4, 5, 6
− (AD ∪ BC ∪ BD ∪ CD)	which equals ABCD	Gate 3, 4, 5, 6

Plus the contribution of six combinations of five gates at a time

+ (AB U AC U AD U BC U BD) which equals ABCD   Gate 1, 2, 3, 4, 5
+ (AB U AC U AD U BC U CD) which equals ABCD   Gate 1, 2, 3, 4, 6
+ (AB U AC U AD U BD U CD) which equals ABCD   Gate 1, 2, 3, 5, 6
+ (AB U AC U BC U BD U CD) which equals ABCD   Gate 1, 2, 4, 5, 6
+ (AB U AD U BC U BD U CD) which equals ABCD   Gate 1, 3, 4, 5, 6
+ (AC U AD U BC U BD U CD) which equals ABCD   Gate 2, 3, 4, 5, 6

Plus the contribution of all six gates at a time which equals ABCD.

Doing the algebra results in:

$$P \text{ (Fire Indication Failure)} = (6 \times 0.01^2) - (8 \times 0.01^3) + (3 \times 0.01^4) = 0.00059203$$

An approximation can be done by using only the first term. In which case the approximation equals:

$$P \text{ (Fire Indication Failure)} = 6 \times 0.01^2 = 0.0006$$

Gate Solution Method

Fault trees may be solved by simply applying the rules of probability at each gate. While this method may appear simple, it is important that the unions be carefully considered so that multiple instances of a given probability are calculated only once. An example of this was presented in Chapter 5 (Figure 5-12).

Consider the case of a pressure switch and a process connection for that pressure switch (Figure C-6. Pressure Sensor Subsystem). If the pressure switch has a PFD of 0.005 and the process connection has a PFD of 0.02, the PFD of the system could be modeled with a fault tree OR gate.

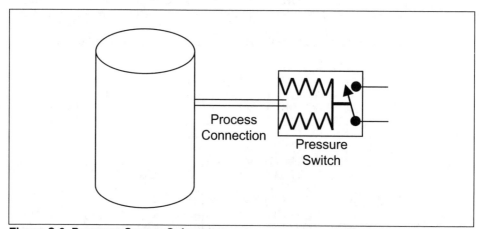

Figure C-6. Pressure Sensor Subsystem

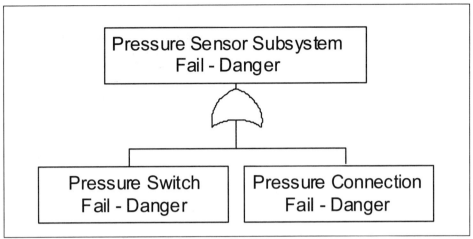

Figure C-7. Pressure Sensor Subsystem Fault Tree

Using a simple gate solution technique the UNION of the probabilities would be $0.02 + 0.005 - (0.02 \times 0.005) = 0.0249$.

NOTE: This is not PFDavg, only PFD. See next section.

Assume that the designer felt this probability of fail-danger to be too high and wanted a second redundant pressure switch. The system is designed to trip if either switch indicates a trip (1oo2 architecture). One might assume the fault tree to be an additional AND gate as shown in Figure C-8.

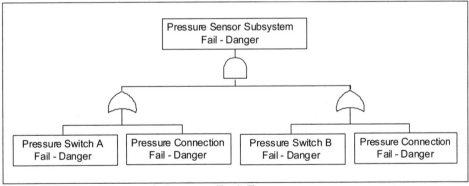

Figure C-8. 1oo2 Pressure Subsystem Fault Tree

When solving this fault tree one must understand if the process connection boxes represent two independent failures each with their own probability or if the two boxes represent one event. A simple gate solution technique that assumes independent events would get the answer $0.0249 \times 0.0249 = 0.00062$. If both boxes are marked identically, often it means they represent one event. In that case the correct answer is $(0.005 \times 0.005) + 0.02 - (0.02 \times 0.005 \times 0.005) = 0.020$. Of course it is recommended that the fault tree be drawn more clearly as is done in Figure C-9.

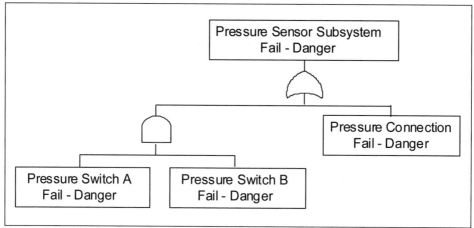

Figure C-9. Clearly Drawn Fault Tree for 1oo2 Pressure Subsystem

PFDavg Calculation

When using fault trees or any other probability combination tool, one must be careful when calculating average probabilities. If PFDavg is the input to any AND gate, the output will be incorrect. It will be incorrect in an optimistic way that may fool the designer into thinking enough safety integrity has been designed when that may not be the case.

The problem is caused by the fact that a product of two integrals does not equal the integral of two products. Remember that PFDavg is obtained by taking an arithmetic average over the time interval T. The average function is defined using an integral.

$$PFDavg = \frac{1}{T}\int_0^T PFD(t)dt$$

If PFDavg is the input to any AND gate, one is obtaining the product of two integrals. As stated above, this is not correct. One must be very careful when obtaining average probabilities.

Use of Fault Trees for PDFavg Calculations

Fault tree analysis can be used to determine the PFDavg of safety instrumented functions. Fault trees are easy to understand, and very complex safety functions can be broken down into individual sections to increase the understanding. The approach commonly used in fault tree analysis is to calculate the PFDavg for each component and then use boolean algebra to calculate the output of the logic gates. This approach is not correct for any AND gate and will lead to optimistic results and potentially unsafe designs. This is further described below for various architectures.

1oo1 Architecture

For a 1oo1 component or system we can use the simplified formula below to calculate the PFDavg, i.e.:

$$\text{PFDavg} = \lambda^D T/2$$

where λ^D is the dangerous undetected fault rate of the component or system and T is the Testing Interval.

1oo2 Architecture

In a 1oo2 architecture, two elements are available to perform the shutdown function and only one is required. This can be represent by an AND gate for probability of dangerous failure (Figure C-10).

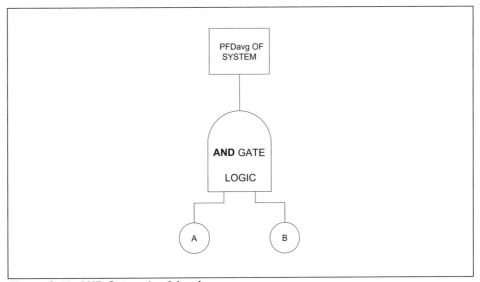

Figure C-10. AND Gate – 1oo2 Logic

Since the PFDavg represents the unavailability of the system when a demand occurs, then for a 1oo2 architecture both components A and B must fail dangerously for a loss of safety function. Both component A AND component B must be failed for the system to fail.

If the PFDavg is used as the input from each device then the PFDavg, using gate probability calculations, for the 1oo2 architecture would be

$$(\lambda^D T/2) \times (\lambda^D T/2) = (\lambda^D)^2 \times T^2/4$$

(this is the result of multiplying average values, i.e. the averaging is done before the logic)

A more accurate equation to calculate the PFDavg for a 1oo2 system is:

$$\text{PFDavg} = (\lambda^D)^2 \times T^2/3$$

This is obtained by averaging after the logic is applied. Calculating the PFDavg before the gate calculations results in lower PFDavg values.

This is because

$$\int_0^T A \times B \text{ is not equal to } \int_0^T A \times \int_0^T B$$

where

$$\text{PFDavg} = \frac{1}{T} \int_0^T PFD(t) dt$$

There is a compensation factor that can be used for rough calculations however.

Note that the PFDavg for a 1oo2 system is $= (\lambda^D)^2 \times T^2/3$

$$= 4/3 \, (\lambda^D T/2)^2$$

$$= 1.33 \, (\text{PDF}_{avg})^2$$

If one were to use the PFDavg as the input to the 1oo2 logic, and the two items had different PFDavg values then the PFDavg for the system will be:

$$= 1.33 \, (\text{PFDavg})_A \times (\text{PFDavg})_B$$

Similarly for a 1oo3 system:

$$\text{PFDavg} = (\lambda_d^3 T^3/4) = 2(\lambda_d T/2)^3$$

$$= 2(\text{PFDavg})_A \times (\text{PFDavg})_B \times (\text{PFDavg})_C$$

for items with different PFDavg values.

As a general rule the PFDavg of a 1ooN system can be determined by multiplying the product of the PFDavg for each system by a factor of:

$$\frac{2^n}{n+1}$$

Note that the "adjustment" factor only works for one level of logic and cannot be used when compound logic is present. It is suggested that under those circumstances, the numerical averaging approach be used.

2oo2 Architecture

For a 2oo2 architecture, if any device or system had failed dangerously when a demand occurs the safety function will be lost i.e. if System A OR system B is unavailable.

The fault tree representing this scenario is as per Figure C-11 below.

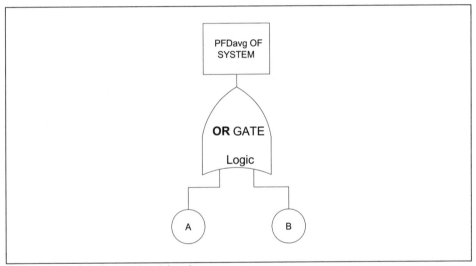

Figure C-11. OR Gate – 2oo2 Logic

If the PFDavg is used as the input from each item then the PFDavg for the 2oo2 architecture would be approximated by:

$$\text{PFDavg} = \lambda^D T/2 + \lambda^D T/2 = \lambda^D T$$

Determination of PFDavg by Averaging the PFD Values Over Operating Time Intervals

The PFDavg can be calculated directly by simply averaging the time dependent PFD values during the operating time interval of interest. The time interval is normally divided into various smaller time intervals and the PFD is calculated for each time interval. The individual values are averaged.

For example, the PFDavg for a device in which the dangerous failure rate was 5.0E-06 failures/hr. and the test interval was every month is calculated. Assuming the manual proof test efficiency was 100%, then we can divide the one month test interval into 30 time intervals (each interval will be 24 hours). Refer to Figure C-12.

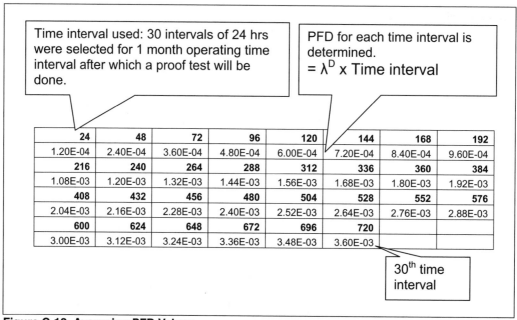

Figure C-12. Averaging PFD Values

The PFD values for the thirty operating time interval increments are added and divided by thirty. The result is:

$$PFDavg = 1.86E\text{-}03$$

Note: For this very simple example the simplified equation the PFDavg value would be:

$$= 5.0E\text{-}06 \times 720/2 = 1.8E\text{-}03$$

Consider an example with a transmitter (PT01) used for a safety function that has the failure rate data as presented in Table C-2.

Table C-2. Transmitter Data

Lambda	%Safe	Automatic Test Diagnostic Coverage CD	Repair Time (RT)	Proof Test Coverage	Lambda DD	Lambda DU E	Lambda DU U
1.50E-06	10%	56%	72	95%	7.50E-07	5.70E-07	3.00E-08

In Table C-2:

- Lambda is the total failure rate of the transmitter
- %Safe is the percentage of total failures that are safe.
- CD is the probability that a dangerous failure will be detected with automatic diagnostics.
- RT is the time to restore the system after a failure occurs.
- The proof test coverage represents the efficiency of the test.
- Lambda DD is the failure rate detected by the automatic diagnostics.
- Lambda DD E is the failure rate detected by the manual test but not detected by automatic diagnostics.
- Lambda DD U is the failure rate undetected by the manual test and not detected by automatic diagnostics.

Manual proof testing is carried out on a monthly basis with an effectiveness of 95%

A full test and complete overhaul is completed every year when the plant is shutdown.

For the above example operating time interval increments of one week are used for the PFD summation. Table C-3 shows the results of the determination of PFDavg by averaging the PFD values over operating time intervals of one week.

Figure C-13 is a plot of the operating time interval versus the PFD value. Notice that because the manual proof test efficiency is 95%, at the end of each one month manual proof test interval the PFD value never goes back to zero. This has a significant impact on the PFD value at the end of the operating time interval.

Table C-3. PFD Values for Transmitter Example

Months	TI	LT	DD	DU E	DU U	DD + DU E + DU U
	182.5	182.5	0.000054	0.000104025	0.000005475	0.0001635
	365	365	0.000054	0.00020805	0.00001095	0.000273
	547.5	547.5	0.000054	0.000312075	0.000016425	0.0003825
1	730	730	0.000054	0.0004161	0.0000219	0.000492
	182.5	912.5	0.000054	0.000104025	0.000027375	0.0001854
	365	1095	0.000054	0.00020805	0.00003285	0.0002949
	547.5	1277.5	0.000054	0.000312075	0.000038325	0.0004044
2	730	1460	0.000054	0.0004161	0.0000438	0.0005139
	182.5	1642.5	0.000054	0.000104025	0.000049275	0.0002073
	365	1825	0.000054	0.00020805	0.00005475	0.0003168
	547.5	2007.5	0.000054	0.000312075	0.000060225	0.0004263
3	730	2190	0.000054	0.0004161	6.57E-05	0.0005358
	182.5	2372.5	0.000054	0.000104025	7.1175E-05	0.0002292
	365	2555	0.000054	0.00020805	7.665E-05	0.0003387
	547.5	2737.5	0.000054	0.000312075	8.2125E-05	0.0004482
4	730	2920	0.000054	0.0004161	8.76E-05	0.0005577
	182.5	3102.5	0.000054	0.000104025	9.3075E-05	0.0002511
	365	3285	0.000054	0.00020805	9.855E-05	0.0003606
	547.5	3467.5	0.000054	0.000312075	0.000104025	0.0004701
5	730	3650	0.000054	0.0004161	0.0001095	0.0005796
	182.5	3832.5	0.000054	0.000104025	0.000114975	0.000273
	365	4015	0.000054	0.00020805	0.00012045	0.0003825
	547.5	4197.5	0.000054	0.000312075	0.000125925	0.000492
6	730	4380	0.000054	0.0004161	0.0001314	0.0006015
	182.5	4562.5	0.000054	0.000104025	0.000136875	0.0002949
	365	4745	0.000054	0.00020805	0.00014235	0.0004044
	547.5	4927.5	0.000054	0.000312075	0.000147825	0.0005139
7	730	5110	0.000054	0.0004161	0.0001533	0.0006234
	182.5	5292.5	0.000054	0.000104025	0.000158775	0.0003168
	365	5475	0.000054	0.00020805	0.00016425	0.0004263
	547.5	5657.5	0.000054	0.000312075	0.000169725	0.0005358
8	730	5840	0.000054	0.0004161	0.0001752	0.0006453
	182.5	6022.5	0.000054	0.000104025	0.000180675	0.0003387
	365	6205	0.000054	0.00020805	0.00018615	0.0004482
	547.5	6387.5	0.000054	0.000312075	0.000191625	0.0005577
9	730	6570	0.000054	0.0004161	0.0001971	0.0006672
	182.5	6752.5	0.000054	0.000104025	0.000202575	0.0003606
	365	6935	0.000054	0.00020805	0.00020805	0.0004701
	547.5	7117.5	0.000054	0.000312075	0.000213525	0.0005796
10	730	7300	0.000054	0.0004161	0.000219	0.0006891
	182.5	7482.5	0.000054	0.000104025	0.000224475	0.0003825
	365	7665	0.000054	0.00020805	0.00022995	0.000492
	547.5	7847.5	0.000054	0.000312075	0.000235425	0.0006015
11	730	8030	0.000054	0.0004161	0.0002409	0.000711
	182.5	8212.5	0.000054	0.000104025	0.000246375	0.0004044
	365	8395	0.000054	0.00020805	0.00025185	0.0005139
	547.5	8577.5	0.000054	0.000312075	0.000257325	0.0006234
12	730	8760	0.000054	0.0004161	0.0002628	0.0007329
				PFDavg		**0.000439053**

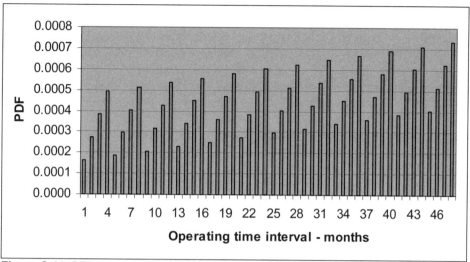

Figure C-13. PFD as a Function of Operating Time Interval

REFERENCES AND BIBLIOGRAPHY

1. Billinton, Roy and Allan, R.N., *Reliability Evaluation of Engineering Systems: Concepts and Techniques*, NY: New York, Plenum Press, 1983.

Appendix D
Markov Models

Introduction

A set of modeling tools based around Markov models can be effectively used to solve a wide variety of reliability and safety problems. Markov models work well as these are stochastic processes, processes where outcomes cannot be accurately predicted but probabilities of outcomes can be obtained.

A Markov system is defined as a "memory-less" system where the probability of moving from one state to another is dependent only upon the current state and not past history of getting to the state. This is the primary characteristic of a Markov model. Markov models are well suited to problems where a state naturally indicates the situation of interest. In some models (characteristic of reliability and safety models) a variable follows a sequence of states. These problems are called Markov chains.

Markov models can deal with a number of complex issues found in the probabilistic modeling of reliability and safety. The models can show system success versus system failure (Figure D-1).

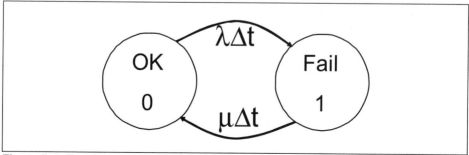

Figure D-1. Two State Model - Success and Failure

Markov models can show redundancy with different levels of redundant components. Figure D-2 shows a system with two subsystems where only one is required for successful system operation. All failures are immediately recognized and the repair probability is modeled as a constant.

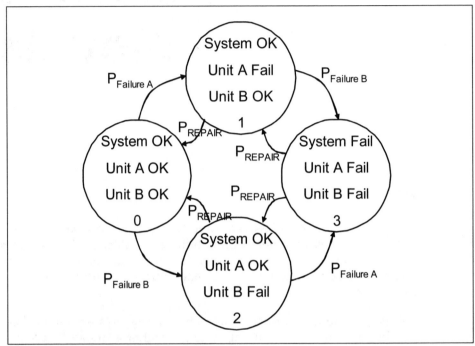

Figure D-2. Redundant Dual System

If both units in a dual redundant system are identical (or close enough so that we do not care which one fails), a model like the one shown in Figure D-2 can be simplified to show only the number of failed units in each state as shown in Figure D-3.

Markov models can show multiple failure states on one drawing (Figure D-4).

Markov Model Types

Markov models have been constructed for four fundamental classes of problems. The models account for two random variables – state and time. The four classes account for continuous states or discrete states and continuous time or discrete time.

Shooman (Ref. 1) presents the example of a shoe box with two partitions such that three compartments exist (Figure D-5). A ping-pong ball is placed in one compartment and the box is tapped on the bottom periodically. At every tap, the ping-pong bounces up. There is a

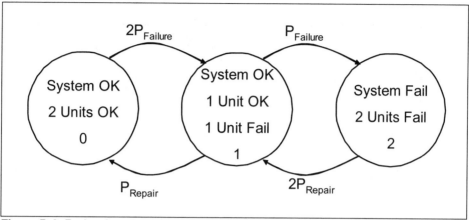

Figure D-3. Redundant Dual System with Identical Units

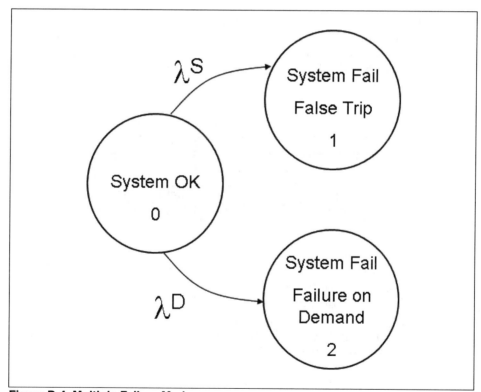

Figure D-4. Multiple Failure Modes

probability that the ball will move to a new compartment or return to the existing compartment. This is an example of a discrete time, discrete space Markov model. If the partitions were removed and the point where the ball landed was recorded, the model would be a discrete time, continuous state model.

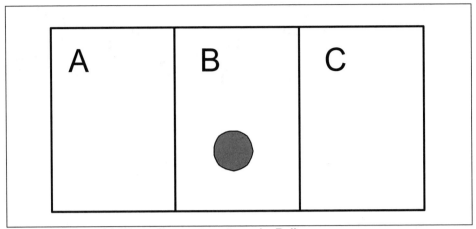

Figure D-5. Box with Three Compartments and a Ball

The continuous state problem is rarely seen in the field of reliability engineering. Both continuous time and discrete time problems are used. Continuous time models typically have analytical solutions. Discrete time models typically have numerical solutions. The model types are listed in Table D-1.

Table D-1. Markov Model Types

State Space	Time	Application
Continuous	Continuous	Not in Reliability Analysis
Continuous	Discrete	Not in Reliability Analysis
Discrete	Continuous	Analytical Solutions
Discrete	Discrete	Numerical Solutions

In addition to the four Markov model types due to space and time variables, a Markov model may have certain additional attributes. A model may be homogenous (often called stationary) or non-homogeneous. A homogeneous model has constant probabilities as a function of time. A non-homogeneous model has probabilities that change as a function of time.

A model may be ergotic (sometimes called "regular") or non-ergotic (sometimes called "absorbing"). In an ergotic model, every state can be reached from any other state either directly or indirectly. In a non-ergotic model one or more states will be absorbing. When the model hits an absorbing state, it stays in that state.

Figures D-1, D-2 and D-3 are examples of ergotic (regular) Markov models. Figure D-4 has an absorbing state (state 2) and is therefore non-ergotic (absorbing).

Markov Model Solution Techniques

Different solution techniques are available for the different model types with different levels of complexity, different assumptions and different effectiveness. The solution techniques available depend on model attributes. A chart of solution techniques is shown in Table D-2 for the discrete state models commonly used in reliability and safety analysis.

Table D-2. Markov Model Solution Techniques

	Ergotic (Regular) Homogeneous	Non-ergotic (Absorbing) Homogeneous	Ergotic (Regular) Non-homogeneous	Non-ergotic (Absorbing) Non-homogeneous
Steady state probability solution via linear equations	Steady State Solution	Not Applicable	Not Applicable	Not Applicable
Time dependent analyticial solutions via differential equations	Continuous Time Solution	Continuous Time Solution	Not Applicable	Not Applicable
Numerical solution for state probability via matrix multiplication	Discrete Time Solution	Discrete Time Solution	Discrete Time Solution	Discrete Time Solution
Analytical / Numerical solution to mean time to first failure via matrix subtraction and inversion	Discrete Time Solution	Discrete Time Solution	Not Applicable	Not Applicable

The Transition Matrix

Any discrete state Markov model may be represented by a square matrix showing the probabilities of moving from one state to another. Such a matrix is called the "transition matrix." This matrix shows the probability of staying in a state as well.

$$P = \begin{bmatrix} 1 - 2P_{Failure} & 2P_{Failure} & 0 \\ P_{Re\,pair} & 1 - (P_{Failure} + P_{Re\,pair}) & P_{Failure} \\ 0 & 2P_{Re\,pair} & 1 - 2P_{Re\,pair} \end{bmatrix}$$

The transition matrix for Figure D-3 is shown above. Notice that a transition matrix is square and sized according to the number of states in the Markov model. Each row in a transition matrix sums to zero.

Notice that the entry in the top left cell is not shown on an arc in Figure D-3. This entry is the probability that the model will remain in state 0 when it is already in state 0. This is not shown on Markov model drawing purely by convention. It is most certainly correct if a modeler wishes to show such transitions on the drawing.

Steady State Probability Solutions

One simple solution technique used for "regular" homogeneous Markov models solves for steady-state probability of being in each state. A "regular" Markov model has no absorbing states. The model may move from any state to any other state, directly or indirectly. All states have at least one exit arc. The steady-state probability solution technique is appropriate and valid only for ergotic (regular) homogeneous (constant probability) Markov models.

> **EXAMPLE D-1**
>
> **Problem:** A shoebox (Figure D-5) is tapped on the bottom every second. A ping-pong ball bounces up and either remains in the compartment it was in or moves to another compartment. If the ball was in compartment A, the probability that it will stay in compartment A is 0.6. If the ball was in compartment A, it will move to compartment B with a probability of 0.4. If the ball was in compartment B before the tap, it will move to compartment A with a probability of 0.3. From compartment B it will move to compartment C with a probability of 0.3. If the ball was in compartment C before the tap it will move to compartment B with a probability of 0.4. The probabilities are constant with time. Create a Markov model and solve it for steady-state probability of being in each compartment.
>
> **Solution:** This problem can be solved with an ergotic, homogeneous Markov model. The state diagram is shown in Figure D-6.

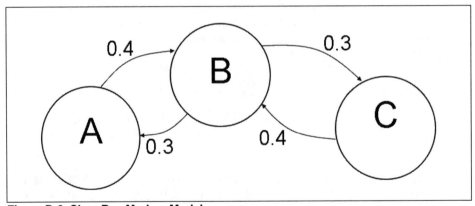

Figure D-6. Shoe Box Markov Model

The transition matrix for this model is:

$$P = \begin{bmatrix} 0.6 & 0.4 & 0 \\ 0.3 & 0.4 & 0.3 \\ 0 & 0.6 & 0.4 \end{bmatrix}$$

The limiting state probabilities are calculated by multiplying a row matrix times the transition matrix and setting that equal to the row matrix. This equation defines the situation when steady-state probabilities are reached.

$$\begin{bmatrix} S_0^L & S_1^L & S_2^L \end{bmatrix} \begin{bmatrix} 0.6 & 0.4 & 0 \\ 0.3 & 0.4 & 0.3 \\ 0 & 0.4 & 0.6 \end{bmatrix} = \begin{bmatrix} S_0^L & S_1^L & S_2^L \end{bmatrix}$$

Multiplying the row matrix times P, three algebraic relationships are obtained:

$$S_0^L = 0.6 S_0^L + 0.3 S_1^L$$
$$S_1^L = 0.4 S_0^L + 0.4 S_1^L + 0.4 S_2^L$$
$$S_2^L = 0.3 S_1^L + 0.6 S_2^L$$

Discarding one of these equations (the equation set provides redundant information) and using the fact that all probabilities sum to one:

$$S_0^L + S_1^L + S_2^L = 1$$

allows a solution for limiting state probabilities. With a little algebraic manipulation of the top line:

$$S_0^L = 0.75 S_1^L$$

Manipulating the third line,

$$S_2^L = 0.75 S_1^L$$

Substituting into the "sum equals one" equation,

$$0.75 S_1^L + S_1^L + 0.75 S_1^L = 1$$

Summing,

$$2.5 S_1^L = 1$$

Finishing the problem, the results are:

$$S_0^L = 0.3$$
$$S_1^L = 0.4$$
$$S_2^L = 0.3$$

Over a long period of time the ball will be in compartment A 30% of the time, compartment B 40% of the time and compartment C 30% of the time.

Repairable Component

Availability and unavailability of a single component can be derived using steady-state probability techniques. Consider the example of a single repairable component, shown in Figure D-1. This model is ergotic and homogeneous. The P matrix for this model is

$$P = \begin{bmatrix} 1-\lambda & \lambda \\ \mu & 1-\mu \end{bmatrix} \tag{D-1}$$

NOTE: The probabilities in a Markov model showing failure rates (λ) and repair rates (μ) are often shown without the "delta t" multiplication. While it is realized that this can appear confusing at first, the simplification in notation helps clarity for most experienced modelers. The transition matrix above shows this simplified notation.

Limiting state probabilities are obtained by multiplying the matrices:

$$\begin{bmatrix} S_0^L & S_1^L \end{bmatrix} \begin{bmatrix} 1-\lambda & \lambda \\ \mu & 1-\mu \end{bmatrix} = \begin{bmatrix} S_0^L & S_1^L \end{bmatrix}$$

This yields:

$$(1-\lambda)S_0^L + \mu S_1^L = S_0^L$$

and

$$\lambda S_0^L + (1-\mu)S_1^L = S_1^L$$

Algebraic manipulation of both equations yield the same result:

$$S_1^L = \frac{\lambda}{\mu} S_0^L$$

Substituting the above into the relation

$$S_0^L + S_1^L = 1$$

yields:

$$S_0^L + \frac{\lambda}{\mu} S_0^L = 1$$

Solving and noting that state 0 is the only success state,

$$A(s) = S_0^L = \frac{\mu}{\lambda + \mu} \qquad (D-2)$$

State 1 is the failure state, therefore unavailability equals:

$$U(s) = S_1^L = 1 - S_0^L = \frac{\lambda}{\lambda + \mu} \qquad (D-3)$$

These results can converted into one of the most well known equations in reliability engineering. Remember that for a single component with a constant failure rate:

$$MTTF = \frac{1}{\lambda}$$

Since a constant repair rate is assumed, MTTR is defined as:

$$MTTR = \frac{1}{\mu}$$

Substituting into Equation D-2, the well known equation

$$A(s) = \frac{MTTF}{MTTF + MTTR} \qquad (D-4)$$

is obtained. It should be remembered that this equation applies to a single component with constant failure rates and constant repair rates.

The steady-state solution technique is useful for many situations. However, it is not appropriate for situations where the probability of moving from state to state is not constant (a non-homogeneous Markov model). It is also not appropriate for absorbing Markov models. This solution technique is not appropriate for safety instrumented functions where many failures are not detected until a periodic inspection and repair is performed. In the case of failures detected by a non-constant inspection and test process, the probability of repair is not constant. It is zero for most time periods. Do not use steady-state techniques to model repair processes with inspection and test.

Time Dependent Analytical Solutions

When the size of the time increment in a discrete Markov model is taken to the limit, the time increment is near zero. The limit as the time increment (delta t) goes to zero is labeled with the letters *dt*. At the limit, we have achieved "continuous" time.

Using continuous time Markov models, analytical solutions for time-dependent state probabilities can be obtained for homogeneous Markov models. As time increments approach zero, the notation Sn(t) is used to indicate time-dependent probability for state n.

Non-repairable Component

An analytical solution for state probabilities for a single non-repairable component (Figure D-7) can be developed by using a little logic and a little calculus. This is a homogeneous, non-ergotic model. A continuous time solution technique will be shown.

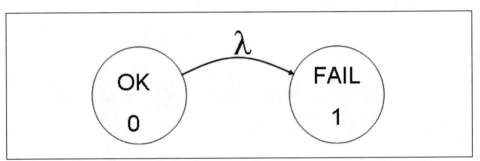

Figure D-7. Single Non-repairable Component

Assume the model starts in state 0 at time t = 0. The model will be in state 0 during the next instant (time = t + delta t) only if it stays in state 0. This can be expressed mathematically as

$$S_0(t + \Delta t) = S_0(t)(1 - \lambda \Delta t) \tag{D-5}$$

This can be rearranged as

$$S_0(t + \Delta t) - S_0(t) = -\lambda S_0(t) \Delta t$$

Dividing both sides by delta t:

$$\frac{S_0(t + \Delta t) - S_0(t)}{\Delta t} = -\lambda S_0(t)$$

The left side of Equation D-5 is the deviation with respect to time. Taking the limit as delta t goes to zero results in:

$$\frac{dS_0(t)}{dt} = -\lambda S_0(t) \tag{D-6}$$

Using a similar process:

$$\frac{dS_1(t)}{dt} = \lambda S_0(t) \tag{D-7}$$

Equations D-6 and D-7 are first order differential equations with constant coefficients. One of the easiest ways to solve such equations is to use a Laplace Transform to convert from the time domain (t) to the frequency domain (s). Taking the Laplace Transforms:

$$sS_0(s) - S_0(0) = -\lambda S_0(s)$$

and

$$sS_1(s) - S_1(0) = \lambda S_0(s)$$

Since the system starts in state 0, substitute $S_0(0) = 1$ and $S_1(0) = 0$. This results in:

$$sS_0(s) - 1 = -\lambda S_0(s) \tag{D-8}$$

and

$$sS_1(s) = \lambda S_0(s) \tag{D-9}$$

Rearranging Equation D-8:

$$(s + \lambda) S_0(s) = 1$$

Therefore:

$$S_0(s) = \frac{1}{s + \lambda} \tag{D-10}$$

Substituting Equation D-10 into Equation D-9 and solving:

$$s S_1(s) = \frac{\lambda}{s + \lambda}$$

Re-arranging:

$$S_1(s) = \frac{\lambda}{s(s+\lambda)}$$

Taking the inverse transform results in:

$$S_0(t) = e^{-\lambda t} \qquad (D\text{-}11)$$

and

$$S_1(t) = 1 - e^{-\lambda t} \qquad (D\text{-}12)$$

Since state 0 is the success state, reliability is equal to $S_0(t)$ and is given by Equation D-11. Unreliability is equal to $S_1(t)$ and is given by Equation D-12. This result is identical to the result obtained when a component has an exponential probability of failure. Thus, the Markov model solution verifies the clear relationship between the constant failure rate and the exponential probability of failure over an interval of time.

Repairable Component

A single repairable component has the Markov model of Figure D-1. To develop time-dependent solutions for the state probabilities, assume that the model starts in state 0. To stay in state 0 during the next instant (time = t + delta t) one of two situations must occur. In situation one, the model must be in state 0 at time t and stay there. In situation two, the model must be in state 1 at time t and move to state 0 during delta t. Mathematically, this is written:

$$S_0(t + \Delta t) = S_0(t)(1 - \lambda \Delta t) + S_1(t)(\mu \Delta t)$$

This can be rearranged as:

$$S_0(t + \Delta t) - S_0(t) = -[\lambda S_0(t) + \mu S_1(t)]\Delta t$$

Dividing both sides of the equation by delta t and taking the limit as delta t goes to zero results in:

$$\frac{d S_0(t)}{dt} = -\lambda S_0(t) + \mu S_1(t) \qquad (D\text{-}13)$$

In a similar manner:

$$\frac{dS_1(t)}{dt} = \lambda S_0(t) - \mu S_1(t) \tag{D-14}$$

Equations D-13 and D-14 are first order differential equations. Again, using the Laplace Transform solution method:

$$sS_0(s) - S_0(0) = -\lambda S_0(s) + \mu S_1(s)$$

Rearranging:

$$(s+\lambda)S_0(s) = \mu S_1(s) + S_0(0)$$

With further algebra:

$$S_0(s) = \frac{\mu}{s+\lambda} S_1(s) + \frac{1}{s+\lambda} S_0(0) \tag{D-15}$$

In a similar manner:

$$S_1(s) = \frac{\lambda}{s+\mu} S_0(s) + \frac{1}{s+\mu} S_1(0) \tag{D-16}$$

Substituting Equation D-15 into Equation D-16:

$$S_1(s) = \frac{\lambda}{s+\mu}\frac{\mu}{s+\lambda} S_1(s) + \frac{\lambda}{s+\mu}\frac{1}{s+\lambda} S_0(0) + \frac{1}{s+\mu} S_1(0)$$

Collecting the terms in a different form:

$$\left(1 - \frac{\lambda}{s+\mu}\frac{\mu}{s+\lambda}\right) S_1(s) = \frac{1}{s+\mu}\left[S_1(0) + \frac{\lambda}{s+\lambda} S_0(0)\right]$$

Creating a common denominator for the left half of the equation yields:

$$\frac{(s+\mu)(s+\lambda) - \lambda\mu}{(s+\mu)(s+\lambda)} S_1(s) = \frac{1}{s+\mu} S_1(0) + \frac{\lambda}{s+\lambda} S_0(0)$$

If both sides of the equation are divided by the first term, the $S_1(s)$ term is isolated.

$$S_1(s) = \frac{(s+\mu)(s+\lambda)}{(s+\mu)(s+\lambda) - \lambda\mu} \frac{1}{s+\mu} S_1(0) + \frac{\lambda}{s+\lambda} S_0(0)$$

Multiplying the denominator of the first term and canceling out equal terms:

$$S_1(s) = \frac{1}{s(s+\lambda+\mu)}\left[(s+\lambda)S_1(0) + \lambda S_0(0)\right] \qquad \text{(D-17)}$$

To move further with the solution, we must arrange Equation D-17 into a form that will allow an inverse transform. A partial fraction expansion of $S_1(s)$ where:

$$S_1(s) = \frac{A}{s} + \frac{B}{s+\lambda+\mu} \qquad \text{(D-18)}$$

will work. This means that

$$\frac{A}{s} + \frac{B}{s+\lambda+\mu} = \frac{1}{s(s+\lambda+\mu)}\left[(s+\lambda)S_1(0) + \lambda S_0(0)\right] \qquad \text{(D-19)}$$

This can be algebraically manipulated into the form:

$$\left[(s+\lambda)S_1(0) + \lambda S_0(0)\right] = A(s+\lambda+\mu) + B(s)$$

This relation holds true for all values of s. Therefore, to solve for A and B, we should pick a value of s that will simplify the algebra as much as possible. To solve for A, a value of $s = 0$ is the best choice. At $s = 0$,

$$\left[\lambda S_1(0) + \lambda S_0(0)\right] = A(\lambda+\mu)$$

Therefore:

$$A = \frac{\lambda}{\lambda+\mu}\left[S_1(0) + S_0(0)\right] \qquad \text{(D-20)}$$

Solving for B gives the result:

$$s = -(\lambda+\mu)$$

Substituting for s:

$$[-\mu S_1(0) + \lambda S_0(0)] = -B(\lambda+\mu)$$

Rearranging,

$$B = \frac{1}{\lambda+\mu}[\mu S_1(0) - \lambda S_0(0)] \quad \text{(D-21)}$$

Substituting equations D-20 and D-21 into D-18:

$$S_1(s) = \frac{\lambda}{\lambda+\mu}\frac{1}{s}[S_1(0)+S_0(0)] + \frac{1}{\lambda+\mu}\frac{1}{s+\lambda+\mu}[\mu S_1(0) - \lambda S_0(0)]$$

Using a similar method for state 0:

$$S_0(s) = \frac{\mu}{\lambda+\mu}\frac{1}{s}[S_1(0)+S_0(0)] + \frac{1}{\lambda+\mu}\frac{1}{s+\lambda+\mu}[\lambda S_0(0) - \mu S_1(0)]$$

Taking the inverse Laplace Transform:

$$S_0(t) = \frac{\mu}{\lambda+\mu}[S_0(0)+S_1(0)] + \frac{e^{-(\lambda+\mu)t}}{\lambda+\mu}[\lambda S_0(0) - \mu S_1(0)]$$

and

$$S_1(t) = \frac{\lambda}{\lambda+\mu}[S_0(0)+S_1(0)] + \frac{e^{-(\lambda+\mu)t}}{\lambda+\mu}[\mu S_1(0) - \lambda S_0(0)]$$

Since the system always starts in state 0:

$$S_0(0) = 1$$

and

$$S_1(0) = 0$$

Substituting,

$$S_0(t) = \frac{\mu}{\lambda+\mu} + \frac{\lambda e^{-(\lambda+\mu)t}}{\lambda+\mu} \quad \text{(D-22)}$$

and

$$S_1(t) = \frac{\lambda}{\lambda+\mu} - \frac{\lambda e^{-(\lambda+\mu)t}}{\lambda+\mu} \tag{D-23}$$

Equations D-22 and D-23 provide the time-dependent analytical formulas for state probability. In this case, equation D-22 is the formula for availability since state 0 is the success state. If t is set equal to infinity in Equation D-22 and Equation D-23, the second term of each goes to zero. The results are:

$$S_0(\infty) = \frac{\mu}{\lambda+\mu} \tag{D-24}$$

and

$$S_1(\infty) = \frac{\lambda}{\lambda+\mu} \tag{D-25}$$

The steady-state probability is the expected result at infinite time. Thus, Equations D-24 and D-25 provide this information and match with D-2 and D-3.

Multiple Failure State Components

Continuous analytical solutions can be obtained for Markov models with multiple failure states. Figure D-4 shows a single component with two failure modes.

Assume that the model starts in state 0. The model will be in state 0 in the time instant only if it stays in state 0. This can be expressed mathematically as

$$S_0(t+\Delta t) = S_0(t)(1 - \lambda^S \Delta t - \lambda^D \Delta t)$$

This can be rearranged as

$$S_0(t+\Delta t) - S_0(t) = -\lambda^S S_0(t)\Delta t - \lambda^D S_0(t)\Delta t$$

Dividing both sides by delta t:

$$\frac{S_0(t+\Delta t) - S_0(t)}{\Delta t} = -\lambda^S S_0(t) - \lambda^D S_0(t) \tag{D-26}$$

The left side of Equation D-26 is the deviation with respect to time. Taking the limit as delta t goes to zero results in:

$$\frac{dS_0(t)}{dt} = -\lambda^S S_0(t) - \lambda^D S_0(t) \tag{D-27}$$

Using a similar process:

$$\frac{dS_1(t)}{dt} = \lambda^S S_0(t) \tag{D-28}$$

and

$$\frac{dS_2(t)}{dt} = \lambda^D S_0(t) \tag{D-29}$$

Equations D-27, D-28 and D-29 are first order differential equations with constant coefficients. Using a Laplace Transform to convert from the time domain (t) to the frequency domain (s):

$$sS_0(s) - S_0(0) = -\lambda^S S_0(s) - \lambda^D S_0(s)$$

$$sS_1(s) - S_1(0) = \lambda^S S_0(s)$$

and

$$sS_2(s) - S_2(0) = \lambda^D S_0(s)$$

For the initial conditions $S_0(0) = 1$, $S_1(0) = 0$ and $S_2(0) = 0$, the equations reduce to:

$$sS_0(s) - 1 = -\lambda^S S_0(s) - \lambda^D S_0(s) \tag{D-30}$$

$$sS_1(s) = \lambda^S S_0(s) \tag{D-31}$$

and

$$sS_2(s) = \lambda^D S_0(s) \tag{D-32}$$

Rearranging Equation D-30:

$$(s + \lambda^S + \lambda^D) S_0(s) = 1$$

Therefore:

$$S_0(s) = \frac{1}{s + \lambda^S + \lambda^D} \qquad (D\text{-}33)$$

Substituting Equation D-33 into Equation D-31 and D-32 and solving:

$$s\, S_1(s) = \frac{\lambda^S}{s + \lambda^S + \lambda^D}$$

A little algebra gives:

$$S_1(s) = \frac{\lambda^S}{s(s + \lambda^S + \lambda^D)} \qquad (D\text{-}34)$$

and

$$S_2(s) = \frac{\lambda^D}{s(s + \lambda^S + \lambda^D)} \qquad (D\text{-}35)$$

Taking the inverse transform of Equation D-33 results in:

$$S_0(t) = e^{-(\lambda^S + \lambda^D)t} = R(t) \qquad (D\text{-}36)$$

Partial fractions must be used proceed with the inverse transform of Equations D-34 and D-35. Using Equation D-34:

$$S_1(s) = \frac{A}{s} + \frac{B}{s + \lambda + \mu} \qquad (D\text{-}37)$$

This means that:

$$\frac{A}{s} + \frac{B}{s + \lambda^S + \lambda^D} = \frac{\lambda^S}{s(s + \lambda^S + \lambda^D)}$$

This can be algebraically manipulated into the form:

$$\lambda^S = A(s + \lambda^S + \lambda^D) + B(s)$$

This relation holds true for all values of s. Therefore, to solve for A and B, we should pick a value of s that will simplify the algebra as much as possible. To solve for A, a value of $s = 0$ is the best choice. At $s = 0$,

$$\lambda^S = A(\lambda^S + \lambda^D)$$

Therefore:

$$A = \frac{\lambda^S}{\lambda^S + \lambda^D} \qquad \text{(D-38)}$$

Solving for B, choose:

$$s = -(\lambda^S + \lambda^D)$$

Substituting for s:

$$\lambda^S = -B(\lambda^S + \lambda^D)$$

Rearranging,

$$B = \frac{-\lambda^S}{\lambda^S + \lambda^D} \qquad \text{(D-39)}$$

Substituting equations E-38 and E-39 into E-37:

$$S_1(s) = \frac{\lambda^S}{s(\lambda^S + \lambda^D)} - \frac{\lambda^S}{(\lambda^S + \lambda^D)(s + \lambda^S + \lambda^D)}$$

Taking the inverse Laplace Transform:

$$S_1(t) = \frac{\lambda^S}{\lambda^S + \lambda^D} - \frac{\lambda^S e^{-(\lambda^S + \lambda^D)t}}{\lambda^S + \lambda^D}$$

This can be rearranged as:

$$S_1(t) = \frac{\lambda^S}{\lambda^S + \lambda^D}(1 - e^{-(\lambda^S + \lambda^D)t}) \qquad \text{(D-40)}$$

Similarly:

$$S_2(t) = \frac{\lambda^D}{\lambda^S + \lambda^D}(1 - e^{-(\lambda^S + \lambda^D)t}) \qquad \text{(D-41)}$$

Equations D-36, D-40 and D-41 provide the time-dependent analytical formulas for state probability. Figure D-8 shows a plot of probabilities as a function of operating time interval. Note that after a long period of time, failure state probabilities begin to reach steady state values.

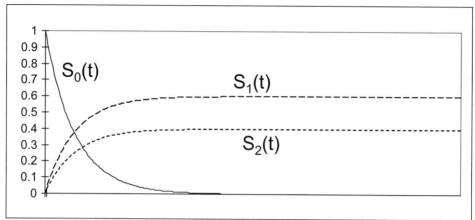

Figure D-8. Multiple Failure State Probabilities

Numerical Solution for State Probability

For discrete time, discrete state Markov models numerical solutions for probability of being in any state can be obtained by simple matrix multiplication. This technique can be used to solve many realistic models, regular or absorbing. The technique may be even used on certain non-homogeneous models to include deterministic events as well as probabilistic events.

To solve for state probabilities, a row matrix indicating starting state probabilities is multiplied by the square transition matrix. Each multiplication represents one discrete time increment.

Consider the Markov model of Example D-1. The transition matrix is multiplied by a row matrix.

$$\begin{bmatrix} S_0^L & S_1^L & S_2^L \end{bmatrix} \begin{bmatrix} 0.6 & 0.4 & 0 \\ 0.3 & 0.4 & 0.3 \\ 0 & 0.4 & 0.6 \end{bmatrix} = \begin{bmatrix} S_0^L & S_1^L & S_2^L \end{bmatrix} \qquad (D\text{-}42)$$

For the numerical solution assume that the problem starts with the ball in box A. Therefore:

1.00	0.00	0.00	0.6	0.4	0
			0.3	0.4	0.3
			0	0.4	0.6

The matrices can easily be entered into a spreadsheet for numerical solution. In Microsoft Excel™ the authors typically set up an operation where the resultant row matrix is located below the input row matrix. In this format time increments descend down the page as shown in Figure D-9.

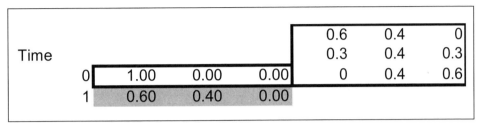

Time						0.6	0.4	0
						0.3	0.4	0.3
	0	1.00	0.00	0.00		0	0.4	0.6
	1	0.60	0.40	0.00				

Figure D-9. Numerical Matrix solution Set Up

This particular resultant matrix is obtained by highlighting the three cell area for the result and entering the command "=MMULT(C5:E5,$F$3:$H$5)" terminated with the keys CTRL-SHIFT-ENTER. If only the ENTER is used, a single value is the result, not the entire row matrix. The process can be easily repeated by copying the resultant cells. This gives state probabilities as a function of discrete time increments. The result is shown in Figure D-10.

Time					0.6	0.4	0
					0.3	0.4	0.3
	0	1.00	0.00	0.00	0	0.4	0.6
	1	0.60	0.40	0.00			
	2	0.48	0.40	0.12			
	3	0.41	0.40	0.19			
	4	0.36	0.40	0.24			
	5	0.34	0.40	0.26			
	6	0.32	0.40	0.28			
	7	0.31	0.40	0.29			
	8	0.31	0.40	0.29			
	9	0.31	0.40	0.29			
	10	0.30	0.40	0.30			
	11	0.30	0.40	0.30			
	12	0.30	0.40	0.30			

Figure D-10. State Probabilities for Each Time Increment

EXAMPLE D-2

Problem: Solve the Markov model of Figure D-4 using a Lambda S of 0.02 failures per hour and a Lambda D of 0.01 failures per hour. The model starts in state 0 where there are no failures.

Solution: The problem can be solved by multiplying a starting row matrix times the transition matrix in a spreadsheet. The numerical result is shown in Figure D-11 and a graph of the results is shown in Figure D-12. Note that the results show the same shape as Figure D-8, the result of the analytical solution.

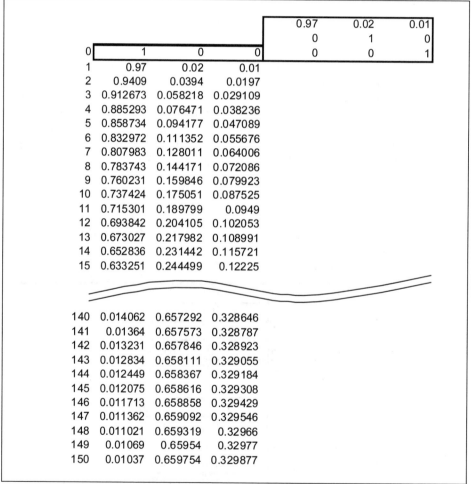

		0.97	0.02	0.01
		0	1	0
		0	0	1
0	1	0	0	
1	0.97	0.02	0.01	
2	0.9409	0.0394	0.0197	
3	0.912673	0.058218	0.029109	
4	0.885293	0.076471	0.038236	
5	0.858734	0.094177	0.047089	
6	0.832972	0.111352	0.055676	
7	0.807983	0.128011	0.064006	
8	0.783743	0.144171	0.072086	
9	0.760231	0.159846	0.079923	
10	0.737424	0.175051	0.087525	
11	0.715301	0.189799	0.0949	
12	0.693842	0.204105	0.102053	
13	0.673027	0.217982	0.108991	
14	0.652836	0.231442	0.115721	
15	0.633251	0.244499	0.12225	
140	0.014062	0.657292	0.328646	
141	0.01364	0.657573	0.328787	
142	0.013231	0.657846	0.328923	
143	0.012834	0.658111	0.329055	
144	0.012449	0.658367	0.329184	
145	0.012075	0.658616	0.329308	
146	0.011713	0.658858	0.329429	
147	0.011362	0.659092	0.329546	
148	0.011021	0.659319	0.32966	
149	0.01069	0.65954	0.32977	
150	0.01037	0.659754	0.329877	

Figure D-11. Numerical Solution for Example D-2

Numerical state probability solution techniques can be used even if probability transitions are deterministic. For example, consider the case when a safety instrumented system is periodically inspected and repaired. The technique is described in detail in Reference 2.

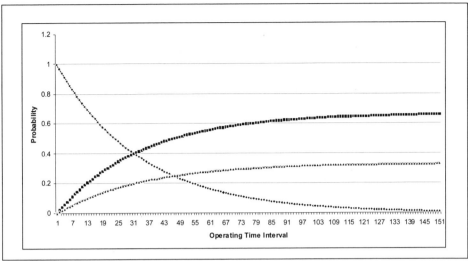

Figure D-12. Plot of State Probabilities for Example D-2

Figure D-13 shows a Markov model with two failure states. Assume that the system being modeled is a single channel (1oo1) system and that all failures in state 1 are immediately recognized. Failures in state 2 are not immediately recognized. They are repaired only after a periodic inspection. If it is assumed that the average inspection and repair time is two hours, the transition probability between state 2 and state 0 can be modeled with by changing the transition matrix as a deterministic function of time. If the system is disabled during the periodic inspection and test, the probability of being in state 2 equals one for that time period. After the inspection and test, the probability of being in state 0 equals one if the inspection, test and repair process is perfect. The transition matrix to be used between inspections is also shown in Figure D-13.

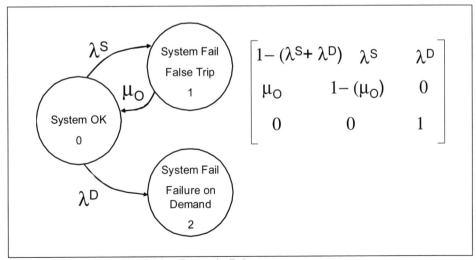

Figure D-13. Markov Model for Example D-3

Appendix D: Markov Model

EXAMPLE D-3

Problem: Solve the Markov model of Figure D-13 using a Lambda S of 0.02 failures per hour and a Lambda D of 0.01 failures per hour. The model starts in state 0 where there are no failures. The system modeled in this example is inspected and repaired to new condition if necessary every 100 hours. Average repair time for all failures is two hours.

Solution: Transition matrix A is used for the time period from 0 to 99 hours. Transition matrix B is used for the time period of 100 to 101 hours. Transition matrix A is used for the time period from 102 to 199 hours. The pattern repeats from this point onward. The solution starts at time = 0 with the model in state 0 as shown in Figure D-14.

Time	State 0	State 1	State 2	0.97	0.02	0.01
				0.5	0.5	0
0	1	0	0	0	0	1

Figure D-14. Example D-3 Starting Matrices

The solution continues by multiplying each row matrix by transition matrix A.

Matrix Solution for Mean Time to First Failure

If the primary metric of interest is "Mean Time To first Failure," MTTF, then a matrix technique can be used either analytically or numerically to find a solution. The technique was described in textbooks as early as the 1950s and well documented in more recent math texts (Ref. 4).

The MTTF for a Markov model can be calculated from the transition matrix. The first step is to create a truncated matrix that contains only the transient success states of the system. This is done by crossing out the rows and the columns of the "absorbing" states or failure states. Using the Markov model from Figure D-15, the truncated matrix, called the Q matrix, is:

$$Q = \begin{matrix} 0.996 & 0.002 & 0.002 \\ 0.1 & 0.899 & 0 \\ 0.1 & 0 & 0.899 \end{matrix} \quad \text{(D-43)}$$

The Q matrix is subtracted from the Identity Matrix, known as I.

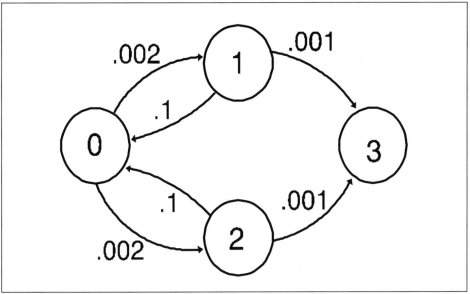

Figure D-15. Markov Model for MTTF Example

$$I - Q = \begin{matrix} 1 & 0 & 0 \\ 0 & 1 & 0 \\ 0 & 0 & 1 \end{matrix} - \begin{matrix} 0.996 & 0.002 & 0.002 \\ 0.1 & 0.899 & 0 \\ 0.1 & 0 & 0.899 \end{matrix}$$

$$= \begin{matrix} 0.004 & -0.002 & -0.002 \\ -0.1 & 0.101 & 0 \\ -0.1 & 0 & 0.101 \end{matrix}$$

(D-44)

Another matrix, called the N matrix, is obtained by inverting the I - Q matrix. Matrix inversion can be done analytically for small matrices, but it is impractical to do for large realistic matrices that represent more complicated systems. A reasonable solution is available, however. Many spreadsheet programs in common use have the ability to numerically invert a matrix. This tool can be used to make quick work of previously time-consuming MTTF calculations. A numerical matrix inversion of the (I - Q) matrix using a spreadsheet provides the following:

$$[I - Q]^{-1} = N = \begin{matrix} 25250 & 500 & 500 \\ 25000 & 504.95 & 495.05 \\ 25000 & 495.05 & 504.95 \end{matrix}$$

(D-45)

The N matrix provides the expected number of time increments that the system dwells in each system success state (a transient state) as a function of starting state. In our example, the top row states the number of time increments per transient state if we start in state 0. The middle row gives the number of time increments if we start in state 1. The bottom row states the number of time increments if we start in state 2. If a system always starts in state 0, we can add the numbers from the top row to get the total number of time increments in all system success states. When this is multiplied by the time increment, we obtain the MTTF when the system starts in state 0. In our example, this number equals 26,250 hours since we used a time increment of one hour. If we started this system in state 1, we would expect the system to fail after 26,000 hours on the average. If we start the system in state 2, we would also expect 26,000 time increments to pass until absorption (26,000 hours until failure).

The technique can be extended to calculate the MTTF for any given failure state. If the metric of interest was mean time to dangerous failure in a multi-failure state Markov model then the model could be modified and the MTTF to a particular state could be calculated (see Ref. 3).

The definition of mean time to first state failure is "The average operating time from start-up in full operating condition (state 0) to failure in the state of interest." This time period includes interruption in failure states other than the one of interest. The method for calculating this metric is:

1. Modify the Markov model to redirect any failure arcs to the failure states not of interest. These are routed to the system success state reached upon repair from the failure state.

2. Eliminate any self loops caused by step 1.

3. Calculate the system MTTF for the revised Markov model

Consider the example from Figure D-16. This Markov model represents a 1oo2 architecture. In order to calculate MTTFS, the mean time to first failure in the Fail-Safe state, the Markov model is modified per the steps above and turns into Figure D-17.

The model can now be solved for MTTF per the established rules provided above. The result is MTTFS, mean time to failure spurious.

REFERENCES AND BIBLIOGRAPHY

1. Shooman, M.L., *Probabilistic Reliability: An Engineering Approach*, Second Edition. Robert E. Krieger Publishing, 1990.

2. Bukowski, J. V. "Modeling and Analyzing the Effects of Periodic Inspection on the Performance of Safety-Critical Systems." *IEEE Transactions of Reliability* (Vol. 50, No. 3). IEEE, September 2001.

3. Bukowski, J. V. and W. M. Goble. "Effects of Maintenance Policies on MTTF of Dangerous Failures in Programmable Electronic Controllers." *ISA Transactions (Vol. 33, No. 1)*. Elsevier, 1994.

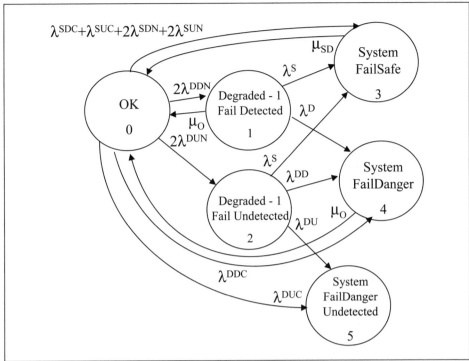

Figure D-16. Multiple Failure State Markov Model

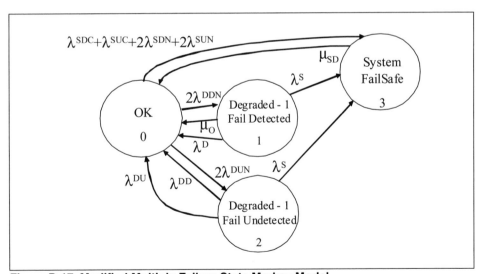

Figure D-17. Modified Multiple Failure State Markov Model

4. Maki, D. P. and M. Thompson. *Mathematical Models and Applications*, Chapter 3. Prentice-Hall, 1973.

Appendix E
Failure Modes Effects and Diagnostic Analysis (FMEDA)

Failure Modes and Effect Analysis (FMEA)

A Failure Modes and Effects Analysis (FMEA) is a systematic technique that is designed to identify problems. It is a "bottom up" method that starts with a detailed list of all components within the system. The overall objective is to identify design flaws, unexpected results, when components of the system fail. A whole system can be analyzed one component at a time.

Alternatively, the system can be hierarchically divided into sub-systems and modules as required. The FMEA can be done on each grouping in the hierarchy. A commonly used reference for the FMEA technique is MIL-STD-1629A (Ref. 1). More recent standards include IEC 60812 (Ref. 2) and SAE J1739 (Ref. 3).

FMEA Procedure

The minimum steps required in the FMEA process are simple:

   A. List all components.

   B. For each component, list all failure modes.

   C. For each component/failure mode, list the effect on the next higher level.

   D. For each component/failure mode, list the severity (i.e., the failure mode of the higher level) of effect.

A FMEA can be very effective in identifying critical failures within a system. One of the primary reasons for doing this is so that the system design can be changed to mitigate, or reduce the likelihood of critical failures. For this reason, the best possible time to do a FMEA is during the design phase of a project. The FMEA should be done while design changes can still be made without disrupting the entire project. Ideally, the completed FMEA will have no critical failures identified. All will have been designed out!

A FMEA also provides important documentation input to the reliability and safety evaluation. The various failure modes in components or modules that must be modeled at the next higher level are identified along with anticipated system effects or failure modes, as we have pointed out earlier.

FMEA Format

A FMEA is documented in a tabular format as shown in Table E-1. Computer spreadsheets are ideal tools for this tabular format. Each column in the table has a specific definition. The following column formats were defined in MIL-STD-1629A (Ref. 1).

Column 1 describes the name of the device under review. Depending on the scope of the FMEA, this could be a component, a module or a unit. Column 2 is available to list the part number or the code number of the device under review. Column 3 describes the function of the component. A good functional description of each component can do an effective job in helping to document system operation.

Column 4 describes the failure modes of the component. One row is typically used for each component failure mode. Examples of component failure modes include fail short, fail open, drift, stuck at one, stuck at zero, etc., for electronic components. Mechanical switch failure modes might include stuck open, stuck closed, contact weld, ground short, etc. Column 5 describes the cause of the failure mode of column four. Generally this is used to list the primary "stress" causing the failure. For example, heat, chemical corrosion, dust, electrical overload, RFI, human operational error, etc.

Column 6 describes how this component failure mode affects the component function of the module (or sub-system). Column 7 lists how this component failure mode affects the next system sub level. In safety evaluations this column is used to indicate safe versus dangerous failures. Depending on the scope of the FMEA, it is possible to consider all levels of the system (component, module, unit or system). Frequently a FMEA at the component level describes the effect at the module level and perhaps the effect at the unit level. Unless the FMEA is being done for a specific system architecture, another FMEA is done for higher levels.

Column 8 is used to list the failure rate of the particular component failure mode. The use of this column is optional when FMEAs are being done for qualitative purposes. When quantitative failure rates are desired and specific data for the application is not available, failure rates and failure mode percentages are available from handbooks (see Ref. 4, 5, 6, and 7).

Lastly column 9 is reserved for comments and relevant information. This area gives the reviewer an opportunity to suggest improvements in design, methods to increase strength of the component (against the perceived stress) or perhaps needed user documentation considerations.

Table E-1. FMEA Tabular Format

1	2	3	4	5	6	7	8	9
Name	Code	Function	Mode	Cause	Effect	Criticality	λ	Remarks
Cool Tank		water storage	leak	corrosion	lost water	dangerous	0.0001	consider design change to detect
			plugged outlet	dirt	no water	dangerous	0.0014	second outlet?
Valve	VALVE1	open for coolant	jam closed	dirt, corr.	no water	dangerous	0.00012	second valve?
			fail open	corr., power	false trip	safe	0.0002	
			coil open	elec. surge	false trip	safe	0.0001	
			coil short	corr., wire	false trip	safe	0.00013	
jacket		path for coolant	leak		none	none	0.000001	
			clog	dirt, corr.	no water	dangerous	0.0001	small flow in normal operation?
drain pipe		path for coolant	clog	dirt, corr.	no water	dangerous	0.00005	
temp. switch	TSW1	sense overtemp	short		no cooling	dangerous	0.0002	two switches?
			open	elec. surge	false trip	safe	0.0002	
power supply	PS1	energy for valve	short	maint. er.	false trip	safe	0.002	
			open	many	false trip	safe	0.004	

Diagnostic Techniques

The self-diagnostic capability of a safety instrumented function is a critical variable for safety integrity verification. Good diagnostics improve both safety and availability. Detection of component failures is done by two different techniques classified as reference or comparison.

Reference diagnostics can be done with a single unit. The coverage factor of reference diagnostics will vary widely with results ranging from 0.0 to 0.99.

Comparison diagnostics require two or more units. The coverage factor depends on implementation but results are generally good with most results ranging from 0.8 to 0.99.

The measure of diagnostic capability is called the "Coverage Factor." It was originally defined as the probability (a number from 0 to 1) that a failure will be detected given that a failure has occurred (Ref. 8). The symbol for coverage factor is "DCF" or "C."

Reference diagnostics take advantage of predetermined characteristics of a successfully operating instrument. Measurements of voltages, currents, signal timing, signal sequence, mechanical position, pressure, sound level, vibration and temperature can be utilized to accurately diagnose component failures. Advanced reference diagnostics include acoustic analysis, partial valve stroke testing, digital signatures and frequency domain analysis.

Reference diagnostics are performed by a single instrument. The notation C^D_1 is used to designate the dangerous diagnostic coverage factor due to single unit reference diagnostics.

Comparison diagnostic techniques depend on comparing data between two or more instruments. The concept is simple. If a failure occurs in the circuitry, processor or memory of one PES unit, there will a difference between data tables in that unit when compared to another unit. Comparisons can be made of input scans, calculation results, output readback scans, and other critical data. Transmitter signals can be compared within limits although this type of testing must be done carefully to ensure that limits are not too tight. The comparison coverage factor will vary since there are tradeoffs between the amount of data compared and the coverage effectiveness. The notation C^D_2 is used to designate dangerous coverage due to comparison diagnostics between two units.

Failure Modes, Effects and Diagnostic Analysis (FMEDA)

Since diagnostics are such a critical variable in the calculations, the ability to measure and evaluate the effectiveness of the diagnostics is important. This is done using an extended failure modes and effects analysis technique (Ref. 9) and verified with fault injection testing (Ref. 10 and 11). The techniques were refined to include multiple failure modes (Ref. 12) and today are commonly used to evaluate diagnostic capability and failure mode split (Ref. 13).

The FMEA approach can be extended to include an evaluation of the diagnostic ability of a circuit, module, unit or system. Additional columns are added to the chart as shown in Table E-2.

Table E-2. FMEDA for Example AC Input Circuit

Failure Modes, Effects and Diagnostic Analysis

1 Name	2 Code	3 Function	4 Mode	5 Cause	6 Effect	7 Criticality λ	8	9 Remarks	10 Det.	11 Diagnostics	12 Mode	13 SD	14 SU	15 DD	16 DU
R1-10K	4555-10	Input threshold	short		Threshold shift	Safe	0.125		0		1	0	0.13	0	0
			open	solder open	open circuit	Safe	0.5		1	lose input pulse	1	0.5	0	0	0
			drift low			Safe	0.01	none until too low	0		1	0	0.01	0	0
			drift high			Safe	0.01	none until too high	1	lose input pulse	1	0.01	0	0	0
R2100K	4555-100	current limit	short		short input	Safe	0.125		1		1	0.13	0	0	0
			open	solder open		Safe	0.5		1	lose input pulse		0	0	0.5	0
			drift low			Safe	0.01	none until too low	0		1	0	0.01	0	0
			drift high			Safe	0.01	none until too high	1	lose input pulse	1	0.01	0	0	0
D1	4200-7	voltage drop	short	surge	overvoltage	Safe	2		1	lose input pulse	1	2	0	0	0
			open		open circuit	Safe	5		1	lose input pulse	1	5	0	0	0
D2	4200-7	voltage drop	short	surge	overvoltage	Safe	2		1	lose input pulse	1	2	0	0	0
			open		open circuit	Safe	5		1	lose input pulse	1	5	0	0	0
OC1	4805-25	isolate	led dim	wear	no light	Safe	28		1	Comp. mismatch	1	28	0	0	0
			tran. short	internal short	read logic 1	Dang.	10		1	Comp. mismatch	0	0	0	10	0
			tran. open		read logic 0	Safe	6		1	Comp. mismatch	1	6	0	0	0
OC2	4805-25	isolate	led dim	wear	no light	Safe	28		1	Comp. mismatch	1	28	0	0	0
			tran. short	internal short	read logic 1	Dang.	10		1	Comp. mismatch	0	0	0	10	0
			tran. open		read logic 0	Safe	6		1	Comp. mismatch	1	6	0	0	0
OC1/OC2			cross channel short		same signal	Dang.	0.01		0		0	0	0	0	0.01
R3-100K	4555-100	filter	short		lose filter	Safe	0.125		0		1	0	0.13	0	0
			open		input float high	Dang.	0.5		1	Comp. mismatch	0	0	0	0.5	0
R4-10K	4555-10	voltage divider	short		read logic 0	Safe	0.125		1	Comp. mismatch	1	0.13	0	0	0
			open		read logic 1	Dang.	0.5		1	Comp. mismatch	0	0	0	0.5	0
R5-100K	4555-100	filter	short		lose filter	Safe	0.125		0		1	0	0.13	0	0
			open		input float high	Dang.	0.5		1	Comp. mismatch	0	0	0	0.5	0
R6-10K	4555-10	voltage divider	short		read logic 0	Safe	0.125		1	Comp. mismatch	1	0.13	0	0	0
			open		read logic 1	Dang.	0.5		1	Comp. mismatch	0	0	0	0.5	0
C1	4350-32	filter	short		read logic 0	Safe	2		1	Comp. mismatch	1	2	0	0	0
			open		lose filter	Safe	0.5		0		1	0	0.5	0	0
C2	4350-32	filter	short		read logic 0	Safe	2		1	Comp. mismatch	1	2	0	0	0
			open		lose filter	Safe	0.5		0		1	0	0.5	0	0
							110.8					86.9	1.4	22.5	0.01

Total Failure Rate	110.8	Safe Coverage 0.9839
Total Safe Failure Rate	88.29	Dang. Coverage 0.9996
Total Dangerous Failure Rate	22.51	
Safe Detected Failure Rate	86.895	
Safe Undetected Failure Rate	1.395	
Dangerous Detected Failure Rate	22.5	
Dangerous Undetected Failure Rate	0.01	

Failures per Billion Hours

The tenth column is an extension to the standard for the purpose of identifying that this component failure is detectable by on-line diagnostics. A number "1" is entered to designate that this failure mode is detectable. A number "0" is entered if the failure mode is not detectable. A number between zero and one can be entered to indicate probability of detection if the diagnostics are not guaranteed to detect the failure.

Column 11 is an extension to the standard used to identify the diagnostic used to detect the failure. The name of the diagnostic should be listed. Perhaps the error code generated or the diagnostic function could also be listed.

Column 12 is used to classify the failure mode with a designation. If only two modes are needed, a "1" and a "0" could be entered. If multiple failure modes are needed, letters can be entered. The failure mode designator can be used in spreadsheets to calculate the various failure rate categories. In Table E-2 only two failure modes are designated, safe and dangerous.

The safe detected failure rate is listed in column 13. This number can be calculated using the previously entered values if a spreadsheet is used for the table. It is obtained by multiplying the failure rate (Column 8) by the failure mode number (Column 12) and the detectability (Column 10).

The safe undetected failure rate is shown in column 14. This number is calculated by multiplying the failure rate (Column 8) by the failure mode number (Column 12) and one minus the detectability (Column 10).

Column 15 lists the dangerous detected failure rate. It is obtained by multiplying the failure rate (Column 8) by one minus the failure mode number (Column 12) and the detectability (Column 10).

Column 16 shows the calculated failure rate of dangerous undetected failures. It is obtained by multiplying the failure rate (Column 8) by one minus the failure mode number (Column 12) and one minus the detectability (Column 10).

Electronic Circuit Design Example

An AC input circuit from a PES is shown in Figure E-1. This circuit was designed for high diagnostic coverage. Very high levels of diagnostics are needed for high safety and high availability systems.

The circuit uses two sets of opto-couplers. The microprocessor that reads the inputs can read both opto-couplers. Under normal operation both readings should be the same. In addition, readings must be taken four times per AC voltage cycle. This allows the microprocessor to read a dynamic signal. When all components are operating properly, a logic 1 is a series of pulses. This circuit design is biased toward fail-safe with a normally energized input sensor. Table E-2 shows the FMEDA for this example.

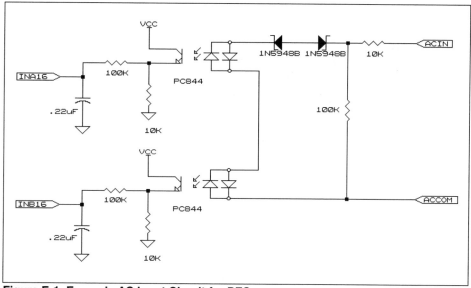

Figure E-1. Example AC Input Circuit for PES

The safe coverage factor for the circuit is calculated by taking the total safe detected failure rate and dividing by the total safe failure rate.

$$C^S = \frac{\sum_{all\ components} \lambda^{SD}_{component\ i}}{\sum_{all\ components} \lambda^{SD}_{component\ i} + \sum_{all\ components} \lambda^{SU}_{component\ i}} \qquad (E-1)$$

The dangerous coverage factor is calculated in a similar way.

$$C^S = \frac{\sum_{all\ components} \lambda^{SD}_{component\ i}}{\sum_{all\ components} \lambda^{SD}_{component\ i} + \sum_{all\ components} \lambda^{SU}_{component\ i}} \qquad (5-2)$$

The analysis indicates that for this circuit, the safe diagnostic coverage factor is 0.96 and the dangerous coverage factor is 0.9996. This is an excellent result.

The FMEDA can also be used as a guide when selecting manual fault injection test cases. If 100% testing is not done, the tests should be done on components with the highest failure rate first. These contribute the most to the coverage factor. Test cases are also chosen when the actual result of the FMEDA is in doubt.

> **EXAMPLE E-1**
>
> **Problem:** A FMEDA shows that the dangerous detected failure rate is 4.84 E-7 failures per hour. The dangerous undetected failure rate is 3.3 E-8 failures per hour. What is the dangerous coverage factor?
>
> **Solution:** The coverage factor is determined using equation E-2.
>
> $$CD = 4.84\ E\text{-}7 / (4.84\ E\text{-}7 + 0.33\ E\text{-}7) = 93.6\%$$

Mechanical Instrument Example

Although the FMEDA technique was developed for electronic circuits, the analysis can be adopted to mechanical instruments (Ref. 14). This technique has been used successfully to estimate the failure rates for various failure mode solenoids, ball valves, control valves, various types of pneumatic and hydraulic actuators, and the mechanical portion of smart valve positioners. Many of the parts must be characterized for the mechanical stress of the particular application but when that is done, the results can be realistic.

The FMEDA method also provides diagnostic coverage factors for the mechanical components when external diagnostic techniques are applied such as partial valve stroke testing.

A portion of a FMEDA for a pneumatic actuator is shown in Table E-3. It can be seen that there is a significant amount of detail and that the systematic approach supplied by the original FMEA technique remains the foundation of the analysis.

Appendix E: FMEDA 311

Table E-3. Partial FMEDA of a Pneumatic Actuator

Item	Part Description	Failure Mode	Effect	Mode	Qty.	Lambda	% distr.
1-10	Housing	Fracture	Torque transmission failure	D	1	5.00E-09	95%
		Deflection	No effect	#	1	5.00E-09	5%
1-20	Housing cover	Fracture	Valve will not move	D	1	5.00E-09	95%
		Deflection	No effect	#	1	5.00E-09	5%
1-30	Guide block assembly	Fracture - piston side power sw	Springforce will cause shut down	S	1	3.00E-08	32%
		Fracture - spring side power sw	Valve will not move	D	1	3.00E-08	32%
		Fracture - middle	Valve will not move	D	1	3.00E-08	32%
		Deflection	No effect	#	1	3.00E-08	5%
1-50	Extension rod assembly	Fracture	Springforce will cause shut down	S	1	5.00E-08	95%
		Deflection	No effect	#	1	5.00E-08	5%
1-60	Extension retainer nut assembly	Loss of Thread	Springforce will cause shut down	S	1	5.00E-08	20%
		Loosen	Springforce will cause shut down	S	1	5.00E-08	80%
1-70	Yoke	Fracture	Valve will not move	D	1	1.00E-07	75%
		Deflection	Valve not fully seated	D	1	1.00E-07	20%
		Wear	Valve not fully seated	D	1	1.00E-07	5%
1-80	Yoke pin	Fracture	Valve will not move	D	1	6.00E-08	95%
		Deflection	Valve not fully seated	D	1	6.00E-08	5%
2-20	Guide bar bearing	Excessive friction	No effect	#	1	3.00E-08	40%
		Excessive play	No effect	#	1	3.00E-08	10%
		Seized	Valve will not move	D	1	3.00E-08	50%
2-25	Yoke pin bearing	Excessive friction	No effect	#	1	3.00E-08	40%
		Excessive play	No effect	#	1	3.00E-08	10%
		Seized	No effect	#	1	3.00E-08	50%
2-30	Yoke/Guide block bushing	Tear	No effect	#	2	3.00E-08	100%
2-40	Yoke bearing	Excessive friction	No effect	#	2	3.00E-08	40%
		Excessive play	No effect	#	2	3.00E-08	10%
		Seized	No effect	#	2	3.00E-08	50%
2-50	O-ring seal	Leak	N/A	#	2		99%
		Complete failure	N/A	#	2		1%
2-80	Rod wiper	N/A	N/A	#	1		100%
2-90	O-ring seal	Leak	N/A	#	2		99%
		Complete failure	N/A	#	2		1%
3-10	Inner end cap	Fracture	Air leak	S	1	2.50E-08	95%
		Deflection	Air leak	S	1	2.50E-08	5%
3-20	Tie bar	Fracture	Valve will not move	D	2	2.50E-08	5%
		Fracture	Release of pressure	S	2	2.50E-08	90%
		Deflection	Valve will not move	D	2	2.50E-08	1%
		Deflection	Release of pressure	S	2	2.50E-08	4%
3-30	Piston	Fracture	Springforce will cause shut down	S	1	2.50E-08	95%
		Deflection	Valve not fully seated	D	1	2.50E-08	5%

The results of the FMEDA on the actuator (Ref. 7) show that the diagnostic coverage factor for dangerous failures is 93.6% for this particular example when partial valve stroke testing is done. It should be noted that the diagnostic coverage result is for a particular design actuator, and the results do not apply to a valve/actuator combination and will likely not apply to any other particular design. Specific analysis must be done for each situation.

FMEDA Limitations

The FMEDA can be very effective but there are limitations. The method shows diagnostics only for known component failure modes. While an extensive body of knowledge exists in databases around the world (Ref. 4 and 7), newly designed components present a risk in that all failure modes may not be known. This can be included in the FMEDA by adding an additional undetected component labeled "unknown" and assigning a failure rate.

As stated above, the technique requires that all failure modes of a component are known. This is all but impossible in complex VLSI integrated circuits like microprocessors. For these circuits an estimate can be made of various failure modes based on manufacturers life test data or failure mode handbooks.

For complex devices with safety critical functionality, more extensive analysis may be justified. A new technique called Random Intelligent Failure Injection Technique (RIFIT) can provide diagnostic coverage via computer simulation of the complex circuit (Ref. 16). Internal faults can be simulated and diagnostics can be measured. The results of a RIFIT analysis can be incorporated into the FMEDA.

REFERENCES AND BIBLIOGRAPHY

1. US MIL-STD-1629A-1984. *Procedures for Performing a Failure Mode Effects and Criticality Analysis.*

2. IEC 60812. Ed. 1.0. 1985. *Analysis Techniques for System Reliability – Procedure for Failure Mode and Effects Analysis (FMEA).*

3. *Potential Failure Mode and Effects Analysis in Design (Design FMEA) and Potential Failure Mode and Effects Analysis in Manufacturing and assembly Processes (Process FMEA) Reference Manual.* Society of Automotive Engineers, 2000.

4. *Failure Mode / Mechanism Distributions.* Reliability Analysis Center, 1997.

5. US MIL-HNBK-217F-1992. Failure Rates for Electronic Components.

6. *Safety Equipment Reliability Handbook.* exida, 2003. (available from ISA)

7. *Component Failure Data Handbook.* exida, 2005.

8. Bouricius, W. G., W. C. Carter, and P. R. Schneider. "Reliability Modeling Techniques for Self-Repairing Systems." *Proceedings of ACM Annual Conference, 1969. Reprinted in Tutorial -- Fault-Tolerant Computing. V. P. Nelson and B. N. Carroll, eds.* IEEE Computer Society Press, 1987.

9. Collett, R. E. and P. W. Bachant. "Integration of BIT Effectiveness with FMECA." *1984 Proceedings of the Annual Reliability and Maintainabiltiy Symposium.* IEEE, 1984.

10. Lasher, R. J. "Integrity Testing of Control Systems." *Control Engineering.* February 1990.

11. Johnson, D. A. "Automatic Fault Insertion." *InTech.* November, 1994 (pp. 42-43).

12. Goble, W. M., J. V. Bukowski, and A. C. Brombacher. "How Diagnostic Coverage Improves Safety in Programmable Electronic Systems." *ISA Transactions (Vol. 36, No. 4).* Elsevier, 1997.

13. Goble, W. M. and Brombacher, A. C., "Using a Failure Modes, Effects and Diagnostic Analysis (FMEDA) to Measure Diagnostic Coverage in Programmable Electronic Systems," *Reliability Engineering & System Safety,* Vol. 66, No. 2, November 1999.

14. Goble, W. M. "Accurate Failure Metrics for Mechanical Instruments." *Proceedings of the IEC61508 Conference* (Augsberg, Germany). RWTUV, January 2003.

15. Brombacher, A. C., Van der Wal, J., Rouvroye, J. L. and Spiker, R., "RIFIT: A Technique to Analyze the Safety of Programmable Safety Systems," *Proceedings of TECH97,* NC: Research Triangle Park, ISA, 1997.

Appendix F
System Architectures

Introduction

This appendix presents a reliability and safety analysis of a number of programmable controller architectures. The architectures chosen represent a majority of those implemented. The architectures are listed in Table F-1.

Table F-1. Architectures

Architecture	Number of units	Output Switches	Objective
1oo1	1	1	Base unit
1oo2	2	2	High Safety
2oo2	2	2	Maintain output
1oo1D	1	2	High Safety
2oo3	3	6	Safety and Availability
2oo2D	2	4	Safety and Availability
1oo2D	2	4	Safety and Availability – biased toward Safety

The architectures modeled in this appendix are the "generic" architectures. Actual commercial implementations may vary. While the architecture concepts are presented with programmable electronic controllers the concepts apply to sensor subsystems and final element subsystems.

This appendix uses material from Chapter 14 of the book *Control System Safety Evaluation and Reliability* by one of the authors (Ref. 1).

Architectures

Programmable Electronic Controllers have been arranged in several configurations in order to achieve various reliability and safety goals. These various arrangements of control system components are referred to as "system architectures." The common system architectures for programmable electronic controllers (PEC) are presented in this appendix along with corresponding fault trees and Markov models.

Single Board PEC

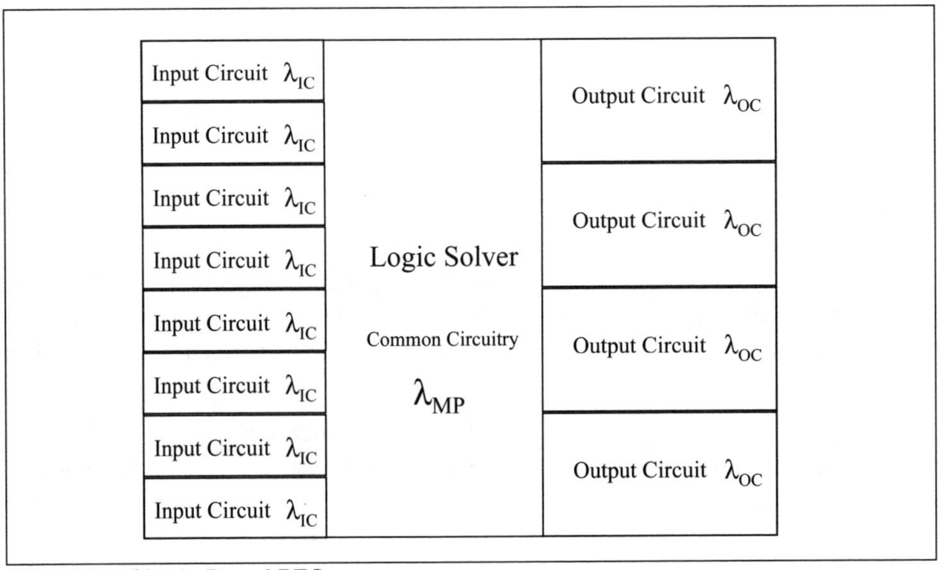

[Diagram showing a Single Board PEC with 8 Input Circuits (λ_{IC}) on the left, a central Logic Solver with Common Circuitry (λ_{MP}), and 4 Output Circuits (λ_{OC}) on the right.]

Figure F-1. Single Board PEC

For comparison purposes a single board PEC with 8 inputs and 4 outputs will be used in a series of examples. Assume that the failure rates are determined by a FMEDA of each circuit. The results of the input circuit analysis indicate the following failure rate categories:

$$\lambda_{IC}^{SD} = 87 \qquad \lambda_{IC}^{SU} = 1 \qquad \lambda_{IC}^{DD} = 22 \qquad \lambda_{IC}^{DU} = 1$$

(All failure rates are in units of FITS, failures / 10^9 hours.)

The results of the output circuit FMEDA are:

$$\lambda_{OC}^{SD} = 85 \qquad \lambda_{OC}^{SU} = 10 \qquad \lambda_{OC}^{DD} = 50 \qquad \lambda_{OC}^{DU} = 1$$

The results of the common circuitry FMEDA are:

$$\lambda_{MP}^{SD} = 8960 \quad \lambda_{MP}^{SU} = 50 \quad \lambda_{MP}^{DD} = 5580 \quad \lambda_{MP}^{DU} = 100$$

It is also assumed that the single board PEC is a series system (the failure of any component is considered a failure of the unit) with constant failure rates. The failure rates may therefore be added to obtain failure rates for the PEC.

$\lambda PEC^{SD} = 8 \times 87 + 4 \times 85 + 8960 = \quad$ 9996 failures per billion hours

$\lambda PEC^{SU} = 8 \times 1 + 4 \times 10 + 50 = \quad$ 98 failures per billion hours

$\lambda PEC^{DD} = 8 \times 22 + 4 \times 50 + 5580 = \quad$ 5956 failures per billion hours

$\lambda PEC^{DU} = 8 \times 1 + 4 \times 1 + 100 = \quad$ 112 failures per billion hours

System Configuration Assumptions

A number of assumptions have been made in these models. It is assumed that:

- Only a single failure can occur on the single board PEC. (One component failure will cause the entire single board PEC to fail.)

- Constant failure rates and repair rates are assumed.

- The models will be based on a de-energize-to-trip system with two failure modes – safe and dangerous.

- Diagnostic test time is much shorter than average repair time.

- The models will account for on-line diagnostic capability and common cause.

- The Markov models will be solved using time dependent solutions not steady-state unavailability.

- Periodic inspection, test and repair will be modeled in the Markov models with a non-constant rate that equals zero except at the test interval so that the model will show state probabilities accurately.

- It is assumed that maintenance policies allow the quick repair of detected dangerous system failures without shutting down the process.

- Perfect inspection and repair are assumed.

- The models also assume that during a repair call, all pieces of equipment are inspected and repaired if failures exist. This may or may not be realistic depending on training and maintenance policy of a particular site.

- The models assume that common cause failures will be the same in all single board PECs.

- When the system fails safely, the failure is self-revealing and is repaired immediately without the need for self-diagnostics and annunciation. This is very realistic since production shutdowns carry a large economic penalty and are repaired quickly. The repair time for such a system failure is not the same as the on-line repair.

- In architectures with diagnostics control output switches (1oo1D, 2oo2D and 1oo2D), it is assumed that detected failures will automatically de-energize outputs (automatic shutdown).

- "Comparison diagnostics" are not assumed.

1oo1 Architecture

A single PEC (Figure F-2) represents a minimum system. No fault tolerance is provided by this system. No failure mode protection is provided. The electronic circuits can fail safely (outputs de-energized, open circuit) or dangerously (outputs frozen or energized, short circuit). Since the effects of on-line diagnostics should be modeled, four failure categories are included:

- DD - dangerous detected,
- DU - dangerous undetected,
- SD - safe detected, and
- SU - safe undetected.

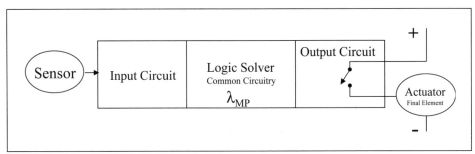

Figure F-2. 1oo1 Architecture

PFD Fault Tree for 1oo1

Figure F-3 shows the fault tree for dangerous failures. The system will fail dangerously if the unit fails dangerous detected (DD) or dangerous undetected (DU).

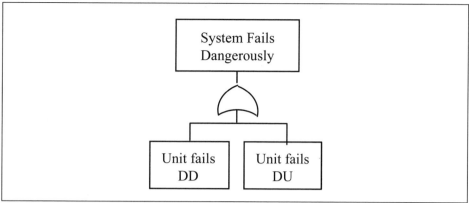

Figure F-3. PFD Fault Tree for the 1oo1 Architecture

Using rough, first order approximation techniques, a simple formula can be generated from the fault tree for the probability of dangerous failure (probability of failure on demand), PFD (assuming perfect periodic test and repair)

$$\text{PFD}_{1oo1} = \lambda^{DD} \times \text{RT} + \lambda^{DU} \times \text{TI} \tag{F-1}$$

where

> RT = average repair time
> TI = test interval for a periodic inspection

It should be pointed out that the approximation techniques are only valid for very small system failure rates.

Since many safety evaluations are done using average probability of failure on demand (PFDavg), the equation for PFDavg should be derived. The average approximation is given by:

$$PFDavg(t) = \frac{1}{t}\int_0^t \left(\lambda^{DD} RT + \lambda^{DU} t'\right) dt'$$

substituting t = TI and integrating

$$PFDavg(TI) = (\lambda^{DD} RT) + \frac{1}{TI}\lambda^{DU}\int_0^{TI}(t') dt'$$

integrating t'

$$PFDavg(TI) = \lambda^{DD} RT + \lambda^{DU}\frac{1}{TI}\left[\frac{t'^2}{2}\right]_0^{TI}$$

when evaluated gives

$$\text{PFDavg}_{1oo1} = \lambda^{DD} \times RT + \lambda^{DU} \times TI/2 \qquad (F-2)$$

for the 1oo1 architecture assuming perfect test and repair.

When imperfect periodic test and repair is considered, the fault tree and the equation get more complicated. The equivalent equation including imperfect test and repair is given by:

$$\text{PFDavg}_{1oo1} = \lambda^{DD} \times RT + C_{PT} \times \lambda^{DU} \times TI/2 + (1 - C_{PT}) \times \lambda^{DU} \times LT/2 \quad (F-3)$$

Where C_{PT} = the diagnostic coverage of the proof test and LT = the lifetime of the instrument. This period is often the time between major overhauls where an instrument is removed from service and completely renewed.

PFS Fault Tree for 1oo1

Figure F-4 shows that the system will fail safely if the unit fails with a safe detected (SD) or safe undetected (SU) failure. NOTE: Remember that the terms "detected" and "undetected" refer to the failures detected by on-line diagnostics, not those revealed by a false trip.

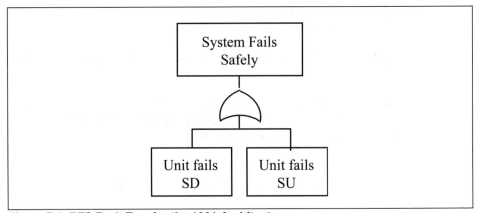

Figure F-4. PFS Fault Tree for the 1oo1 Architecture

The same approximation techniques can be used to generate a formula for probability of failing safely, PFS

$$\text{PFS}_{1oo1} = \lambda^{SD} \times SD + \lambda^{SU} \times SD \qquad (F-4)$$

where

SD = the average time required to restart the process after a shutdown.

Markov Model for 1oo1

The 1oo1 architecture can also be modeled using a Markov model (Figure F-5). In the Markov model for this configuration, state 0 represents the condition where there are no failures. From this state, the controller can reach three other states. State 1 represents the fail-safe condition. In this state, the controller has failed with its outputs de-energized. State 2 represents the fail-danger condition with a detected failure. In this state, the controller has failed with its outputs energized but the failure is detected by diagnostics and can be repaired. The system has also failed dangerously in state 3 but the failure is not detected by on-line diagnostics.

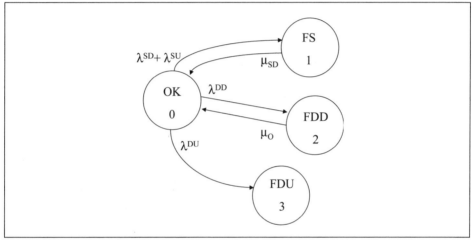

Figure F-5. 1oo1 Markov Model

For the time period before the first periodic inspection, the transition matrix, P, for the 1oo1 system is:

$$P = \begin{bmatrix} 1-(\lambda^S + \lambda^D) & \lambda^{SD} + \lambda^{SU} & \lambda^{DD} & \lambda^{DU} \\ \mu_{SD} & 1-\mu_{SD} & 0 & 0 \\ \mu_O & 0 & 1-\mu_O & 0 \\ 0 & 0 & 0 & 1 \end{bmatrix}$$

To derive the MTTF formula, standard linear algebra techniques can be used (Ref. 1, Chapter 8). As a first step, the transition matrix failure state rows and columns associated with the failure states are truncated. This operation yields the Q matrix.

$$Q = \begin{bmatrix} 1-(\lambda^S + \lambda^D) \end{bmatrix}$$

This is subtracted from the identity matrix:

$$I - Q = 1 - \left[1 - (\lambda^S + \lambda^D)\right] = \lambda^S + \lambda^D$$

The N matrix is obtained from $[I - Q]^{-1}$. In this case:

$$N = \frac{1}{\lambda^S + \lambda^D}$$

Since MTTF is a sum of the row elements of the N matrix for a given starting state and there is only one starting state, it is simply:

$$MTTF = \frac{1}{\lambda^S + \lambda^D} \tag{F-5}$$

Example Calculations for the 1oo1 Architecture

Using the example parameters:

Table F-2. Parameters for Model Comparison

Model Parameters	
λ_{PEC}^{SD}	9996 FITS
λ_{PEC}^{SU}	98 FITS
λ_{PEC}^{DD}	5956 FITS
λ_{PEC}^{DU}	112 FITS
RT	8 Hours
TI	8760 Hours
SD	24 Hours

First, calculations are done using the first order approximation equations derived from the fault tree. The PFD calculation using Equation F-1:

$$PFD_{1oo1} = 0.000005956 \times 8 + 0.000000112 \times 8760 = 0.001029$$

The PFDavg calculation using Equation F-2:

$$PFDavg_{1oo1} = (0.000005956 \times 8 + 0.000000112) \times (8760/2) = 0.000538$$

The PFS calculation using Equation F-4:

$$PFS_{1oo1} = (0.000009996 \times 24) + (0.000000098 \times 24) = 0.000242$$

Calculations are also done using matrix math solutions derived from the Markov model. MTTF (Mean Time to any Failure) for a 1oo1 configuration is calculated using Equation F-5. The total safe failure rate is 10094 FITS. The total dangerous failure rate is 6068 FITS. The total failure rate is 16162 FITS. Using Equation F-5:

$$\text{MTTF} = 1 / 16162 \times 10^{-9} = 61{,}874 \text{ hours}$$

The time dependent probabilities can be calculated by multiplying the P matrix times a row matrix, S. Assuming that the unit is working properly when started, the system starts in state 0 and the S matrix is: S = [1 0 0 0]. First the failure rate and repair rate numbers are substituted into a P matrix. The repair rates are given by

$$\mu_O = \frac{1}{RT} \quad \text{and} \quad \mu_{SD} = \frac{1}{SD}$$

The numeric matrix is:

P	0	1	2	3
0	0.9999838	0.0000101	0.0000060	0.0000001
1	0.0416667	0.9583333	0.0000000	0.0000000
2	0.1250000	0.0000000	0.8750000	0.0000000
3	0.0000000	0.0000000	0.0000000	1.0000000

The PFS at TI = 8760 hours is 0.0002419. The PFD is the sum of state 2 and state 3 probabilities. At TI = 8760 hours, the PFD is 0.001028. The results are summarized in Table F-3.

Table F-3. Results from 1oo1 Calculations

Calculation results for 1oo1 Architecture		
FT Equation	PFD	0.001029
FT Equation	PFDavg	0.000538
FT Equation	PFS	0.000242
Markov Solution	PFD	0.001028
Markov Solution	PFS	0.000242
Markov Solution	MTTF	61,874Hrs

It should be noted that the approximation formulas are solved for a particular value of periodic inspection (TI). The Markov models are solved as a function of operating time interval. When comparing results, the Markov model value for a time interval of TI is used.

1oo2 Architecture

Two controllers can be wired to minimize the effect of dangerous failures. For de-energize-to-trip systems, a series connection of two output circuits requires that both controllers fail in a dangerous manner for the system to fail dangerously. The 1oo2 configuration typically utilizes two independent main processors with their own independent I/O (see Figure F-6). The system offers low probability of failure on demand, but it increases the probability of a fail-safe failure. The "false trip" rate is increased in order to improve the ability of the system to shut down the process.

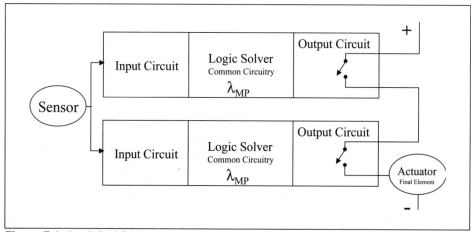

Figure F-6. 1oo2 Architecture

PFD Fault Tree for 1oo2

Figure F-7 shows the PFD Fault Tree for the 1oo2 architecture. The system can fail dangerously if both units fail dangerously due to a common cause failure, detected or undetected. Other than common cause, it can fail dangerously only if both A and B fail dangerously.

A first order approximation for PFD can be derived from the fault tree. The equation for PFD is:

$$PFD_{1oo2} = \lambda^{DUC} \times TI + \lambda^{DDC} \times RT + (\lambda^{DDN} \times RT + \lambda^{DUN} \times TI)^2 \qquad (F-6)$$

The approximation equation for PFDavg derived from the fault tree is:

$$PFDavg_{1oo2} = \lambda^{DUC} \times (TI/2)$$
$$+ \lambda^{DDC} \times RT + (\lambda^{DDN} \times RT)^2$$
$$+ (\lambda^{DDN} \times RT \times \lambda^{DUN} \times TI)^2/2 + (\lambda^{DUN} \times TI)^2/3 \qquad (F-7)$$

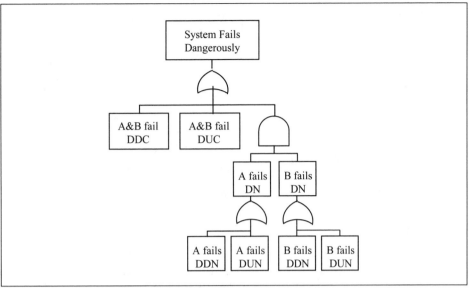

Figure F-7. PFD Fault Tree for 1oo2 Architecture

When imperfect proof test is considered, the equation becomes:

$$\text{PFDavg}_{1oo2} = C_{PT} \times \lambda^{DUC} \times (TI/2) + (1 - C_{PT}) \times \lambda^{DUC} \times (LT/2)$$

$$+ \lambda^{DDC} \times RT + (\lambda^{DDN} \times RT)^2$$

$$+ (\lambda^{DDN} \times RT \times C_{PT} \times \lambda^{DUN} \times TI)^2/2 + (\lambda^{DDN} \times RT \times (1 - C_{PT}) \times \lambda^{DUN} \times LT)^2/2$$

$$+ (C_{PT} \times \lambda^{DUN} \times TI)^2/3 + ((1 - C_{PT}) \times \lambda^{DUN} \times LT)^2/3 \tag{F-8}$$

A comparison of Equation F-7 with Equation F-8 shows that any term from F-7 that contains a TI has a proof test coverage multiplier and that a duplicate term is added with $(1 - C_{PT})$ and LT substituted in Equation F-8.

PFS Fault Tree for 1oo2

Figure F-8 shows the PFS fault tree for the 1oo2 architecture. This shows the tradeoff in the architecture, any safe failure from either unit will cause a false trip.

An approximation equation for PFS can be derived from the fault tree.

$$\text{PFS}_{1oo2} = (\lambda^{SDC} + \lambda^{SUC} + 2\lambda^{SDN} + 2\lambda^{SUN}) \times SD \tag{F-9}$$

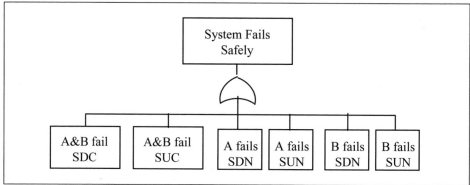

Figure F-8. PFS Fault Tree for 1oo2 Architecture

Markov Model for 1oo2

The Markov model for a 1oo2 single-board system is shown in Figure F-9. Three system success states exist. In state 0, both controllers operate. In state 1 and state 2, one controller has failed with outputs energized. The system is successful because the other controller can still de-energize as required. Since failures in state 1 are detected, an on-line repair rate returns from state 1 to state 0. States 3, 4 and 5 are the system failure states. In state 3, the system has failed with its outputs de-energized. In state 4, the system has failed detected with its outputs energized. An undetected failure with outputs energized has occurred in state 5. Note that only a dangerous common-cause failure will move the system from state 0 to states 4 or 5.

The on-line repair rate from state 4 to state 0 assumes that the repairman will inspect the system and repair all failures when making a service call. If that assumption was not valid, state 4 must be split into two states one with both units failed detected and the other with one detected and one undetected. The state with both detected will repair to state 0. The state with only one detected will repair to state 2. The assumption made for this model does simplify the model and is not significant unless coverage factors are low.

The transition matrix, P, for the 1oo2 system is:

$$P = \begin{bmatrix} 1-(\lambda^{DC}+2\lambda^{DN}+\lambda^{SC}+2\lambda^{SN}) & 2\lambda^{DDN} & 2\lambda^{DUN} & \lambda^{SC}+2\lambda^{SN} & \lambda^{DDC} & \lambda^{DUC} \\ \mu_O & 1-(\lambda^S+\lambda^D+\mu_O) & 0 & \lambda^S & \lambda^D & 0 \\ 0 & 0 & 1-(\lambda^S+\lambda^D) & \lambda^S & \lambda^{DD} & \lambda^{DU} \\ \mu_{SD} & 0 & 0 & 1-\mu_{SD} & 0 & 0 \\ \mu_O & 0 & 0 & 0 & 1-\mu_O & 0 \\ 0 & 0 & 0 & 0 & 0 & 1 \end{bmatrix}$$

Numeric solutions for PFD, PFS, MTTF and other reliability metrics can be obtained from this matrix using a spreadsheet.

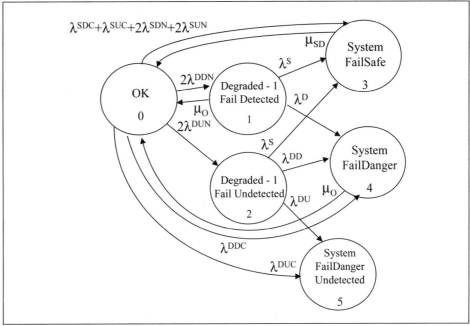

Figure F-9. Markov Model for 1oo2 Architecture

Example Calculations for the 1oo2 Architecture

In the 1oo2 architecture, redundant units are used. This means that common cause must be included in the modeling. Using a Beta factor of 0.02, the model parameters are shown in Table F-4.

Table F-4. Model Parameters Including Common Cause Failure Rates

Model Parameters	
λ_{PEC}^{SDC}	199.9 FITS
λ_{PEC}^{SUC}	2.0 FITS
λ_{PEC}^{SDN}	9796.1 FITS
λ_{PEC}^{SUN}	96.0 FITS
λ_{PEC}^{DDC}	119.1 FITS
λ_{PEC}^{DUC}	2.2 FITS
λ_{PEC}^{DDN}	5836.9 FITS
λ_{PEC}^{DUN}	109.8 FITS
Beta	0.02
RT	8 Hours
TI	8760 Hours
SD	24 Hours

Using equation substitution, the fault tree based approximations can easily be solved. Using a spreadsheet tool, the Markov model solutions are obtained using linear algebra. For example, the numerical P matrix is:

P		0	1	2	3	4	5
	0	0.9999680	0.0000117	0.0000002	0.0000200	0.0000001	0.0000000
	1	0.1250000	0.8749838	0.0000000	0.0000101	0.0000061	0.0000000
	2	0.0000000	0.0000000	0.9999838	0.0000101	0.0000060	0.0000001
	3	0.0416667	0.0000000	0.0000000	0.9583333	0.0000000	0.0000000
	4	0.1250000	0.0000000	0.0000000	0.0000000	0.8750000	0.0000000
	5	0.0000000	0.0000000	0.0000000	0.0000000	0.0000000	1.0000000

The P matrix is truncated to obtain the Q matrix.

Q	0.9999680	0.0000117	0.0000002
	0.1250000	0.8749838	0.0000000
	0.0000000	0.0000000	0.9999838

The Q matrix is subtracted from the identity matrix.

I-Q	0.0000320	-0.0000117	-0.0000002
	-0.1250000	0.1250162	0.0000000
	0.0000000	0.0000000	0.0000162

The I - Q matrix is inverted to obtain the N matrix.

N	49192.00	4.59	668.15
	49185.64	12.59	668.06
	0.00	0.00	61873.53

The MTTF is obtained by adding the row elements of the starting state. Assuming that the system starts in state 0, MTTF = 49192 + 4.6 + 668.2 = 49865 hours.

The PFD and PFS at 8760 hours are calculated by multiplying a starting row matrix (starting in state 0) times the P matrix repeatedly. The results of the calculations for the 1oo2 architecture are shown in Table F-5.

Table F-5. Results from 1oo2 architecture calculations

Calculation results for 1oo2 Architecture		
FT Equation	PFD	0.00002159
FT Equation	PFDavg	0.00001110
FT Equation	PFS	0.00047967
Markov Solution	PFD	0.00002153
Markov Solution	PFS	0.00047898
Markov Solution	MTTF	49,865 Hrs.

2oo2 Architecture

Another dual controller configuration was developed for the situation in which it is undesirable to fail with outputs de-energized. This system is used in energize-to-trip protection systems. The outputs of two controllers are wired in parallel (Figure F-10). If one controller fails with its output de-energized, the other is still capable of energizing the load.

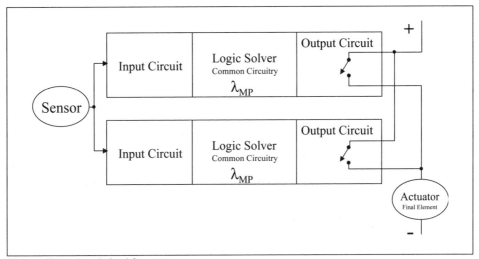

Figure F-10. 2oo2 Architecture

A disadvantage of the configuration is its susceptibility to failures in which the output is energized. If either controller fails with its output energized, the system has failed with output energized. This configuration is not suitable for de-energize to trip protection systems unless each unit is of an inherently fail-safe design.

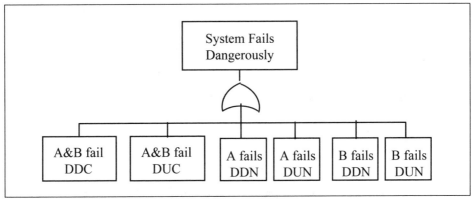

Figure F-11. PFD Fault Tree for the 2oo2 Architecture

PFD Fault Tree for 2oo2

Since the controllers are wired in parallel, any short circuit (dangerous) failure of the components results in a dangerous (outputs energized) failure of the system. This is shown in Figure F-11. The first order approximation equation to solve for PFD is:

$$PFD_{2oo2} = \lambda^{DDC} \times RT + \lambda^{DUC} \times TI + 2\lambda^{DDN} \times RT + 2\lambda^{DUN} \times TI \quad (F\text{-}10)$$

and the equation for $PFDavg_{2oo2}$ derived from the fault tree is:

$$PFDavg_{2oo2} = \lambda^{DDC} \times RT + \lambda^{DUC} \times TI/2 + 2\lambda^{DDN} \times RT + 2\lambda^{DUN} \times TI/2 \quad (F\text{-}11)$$

PFS Fault Tree for 2oo2

This architecture is designed to tolerate an open circuit failure. The fault tree of Figure F-12 shows this. The system will fail open circuit (de-energized, safe) if there is a safe common cause failure. Other than common cause, an open circuit failure on both A and B must occur.

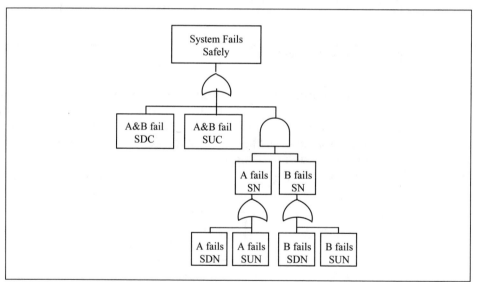

Figure F-12. Fault Tree for the 2oo2 Architecture

The first order approximation equation to solve for PFS is

$$PFS_{2oo2} = \lambda^{SDC} \times SD + \lambda^{SUC} \times SD + 2(\lambda^{SDN} \times RT \times \lambda^{SDN} \times SD + \lambda^{SUN} \times TI \times \lambda^{SDN} \times SD + \lambda^{SDN} \times RT \times \lambda^{SUN} \times SD + \lambda^{SUN} \times TI \times \lambda^{SUN} \times SD) \quad (F\text{-}12)$$

The equation format is not obvious from looking at the fault tree. The equation format results from the fact that there are actually eight ways to get two independent failures of A and B. These are listed in Table F-6.

Table F-6. Permutations of Independent Failure of A and B

First failure	Second failure
A, SDN	then B, SDN
A, SUN	then B, SDN
A, SDN	then B, SUN
A, SUN	then B, SUN
B, SDN	then A, SDN
B, SUN	then A, SDN
B, SDN	then A, SUN
B, SUN	then A, SUN

A more detailed fault tree that utilizes the fault tree symbol for a sequence and function is shown in Figure F-13. This fault tree is much more representative.

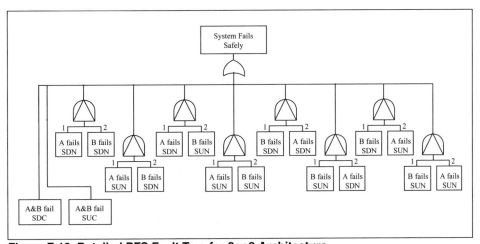

Figure F-13. Detailed PFS Fault Tree for 2oo2 Architecture

Markov Model for 2oo2

The single-board controller Markov model for the 2oo2 architecture is shown in Figure F-14. The system is successful in three states, 0, 1, and 2. The system has failed with outputs de-energized in state 3. The system has failed with outputs energized in states 4 and 5.

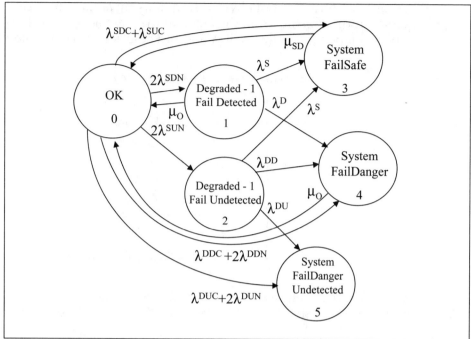

Figure F-14. 2oo2 Markov Model

A comparison inspection of this Markov model to that of the 1oo2 architecture will show a certain symmetry that makes sense given the architectures. The P matrix for this model is:

$$P = \begin{bmatrix} 1-(\lambda^{SC}+2\lambda^{SN}+\lambda^{DC}+2\lambda^{DN}) & 2\lambda^{SDN} & 2\lambda^{SUN} & \lambda^{SC} & \lambda^{DDC}+2\lambda^{DDN} & \lambda^{DUC}+2\lambda^{DUN} \\ \mu_O & 1-(\lambda^S+\lambda^D+\mu_O) & 0 & \lambda^S & \lambda^D & 0 \\ 0 & 0 & 1-(\lambda^S+\lambda^D) & \lambda^S & \lambda^{DD} & \lambda^{DU} \\ \mu_{SD} & 0 & 0 & 1-\mu_{SD} & 0 & 0 \\ \mu_O & 0 & 0 & 0 & 1-\mu_O & 0 \\ 0 & 0 & 0 & 0 & 0 & 1 \end{bmatrix}$$

Example Calculations for the 2oo2 Architecture

Using the example model parameters from Table F-7, solutions can be obtained using both the approximation equations and the Markov model solutions. Substituting numerical values into the P matrix yields:

P	0	1	2	3	4	5
0	0.99996800	0.00001959	0.00000019	0.00000020	0.00001179	0.00000022
1	0.12500000	0.87498384	0.00000000	0.00001009	0.00000607	0.00000000
2	0.00000000	0.00000000	0.99998384	0.00001009	0.00000596	0.00000011
3	0.04166667	0.00000000	0.00000000	0.95833333	0.00000000	0.00000000
4	0.12500000	0.00000000	0.00000000	0.00000000	0.87500000	0.00000000
5	0.00000000	0.00000000	0.00000000	0.00000000	0.00000000	1.00000000

Truncating the rows and columns of the failure states will provide the Q matrix. This matrix is subtracted from the identity matrix to obtain:

```
I-Q         0.00003200   -0.00001959  -0.00000019
           -0.12500000    0.12501616   0.00000000
            0.00000000    0.00000000   0.00001616
```

The I - Q matrix is inverted to obtain the N matrix.

```
N          80572.82      12.63      957.58
           80562.41      20.62      957.46
               0.00       0.00    61873.53
```

Adding the row elements from row 0 provides the answer, MTTF = 81,282 hours.

The PFS and PFD are calculated by multiplying the P matrix by a row matrix S starting with the system in state 0. A set of results is shown in Table F-7. When comparing the results with Table F-7 from the 1oo1 architecture, the PFD values are roughly twice as high. The PFS values are much lower. These are the objectives of the 2oo2 architecture design.

Table F-7. Calculation Results for the 2oo2 Architecture

Calculation results for 2oo2 Architecture		
FT Equation	PFD	0.00203696
FT Equation	PFDavg	0.00106565
FT Equation	PFS	0.00000528
Markov Solution	PFD	0.00203354
Markov Solution	PFS	0.00000524
Markov Solution	MTTF	81,543 Hrs.

1oo1D Architecture

Figure F-15 shows an architecture that uses a single controller channel with diagnostic capability and a second diagnostic channel wired in series to utilize the diagnostic signal to de-energize the output. This differs from the 1oo1 only in that the switch is wired in series with the output to de-energize the output on a diagnostic fault. This system represents an enhancement used for safety applications. Diagnostics allow a detected dangerous failure to be converted into a safe failure. In general, additional failure rates must be included in quantitative analysis to account for the extra diagnostic channel. In systems using external diagnostic control devices (like watchdog timers), additional failure rates for these external devices must be added to the single-board rates.

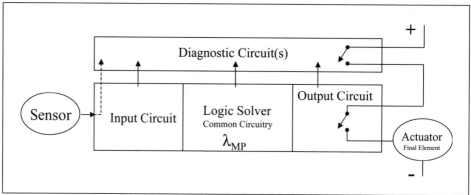

Figure F-15. 1oo1D Architecture

PFD Fault Tree for 1oo1D

The 1oo1D architecture has a second diagnostic channel that will de-energize when failures are detected by the diagnostics. Therefore, the only failures that cause system failure with outputs energized are dangerous undetected failures. The fault tree has only one failure group, DU, as shown in Figure F-16.

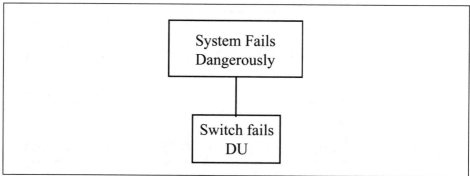

Figure F-16. PFD Fault Tree for the 1oo1D Architecture

The approximation equation derived from the fault tree for PFD is

$$\text{PFD}_{1oo1D} = \lambda^{DU} \times \text{TI} \tag{F-13}$$

The approximate equation for PFDavg is:

$$\text{PFDavg}_{1oo1D} = \lambda^{DU} \times \text{TI}/2 \tag{F-14}$$

PFS Fault Tree for 1oo1D

Figure F-17 shows that a 1oo1D architecture will fail safely if the unit fails with SD, SU or DD failures.

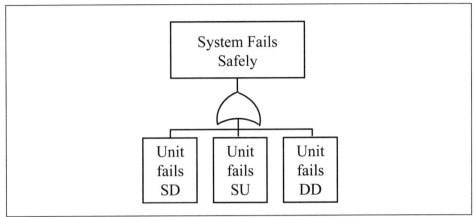

Figure F-17. PFS Fault Tree for the 1oo1D Architecture

The approximation techniques can be used to generate a formula for probability of failing safely from this fault tree.

$$\text{PFS}_{1oo1D} = (\lambda^{SD} + \lambda^{SU} + \lambda^{DD}) \times SD \tag{F-15}$$

where SD = the time required to restart the process after a shutdown.

Markov Model for 1oo1D

The 1oo1D architecture can also be modeled using a Markov model (Figure F-18).

In the Markov model for this configuration, state 0 represents the condition where there are no failures. From this state, the controller can reach two other states. State 1 represents the fail-safe condition. In this state, the controller has failed with its outputs de-energized. The system has failed dangerously in state 3 and the failure is not detected by on-line diagnostics. The Markov model for the 1oo1D is similar to the 1oo1 except that the dangerous detected failures automatically trip the system (go to state 1).

The transition matrix, P, for the 1oo1D system is:

$$P = \begin{bmatrix} 1-(\lambda^{SD} + \lambda^{SU} + \lambda^{DD} + \lambda^{DU}) & (\lambda^{SD} + \lambda^{SU} + \lambda^{DD}) & \lambda^{DU} \\ \mu_{SD} & 1-\mu_{SD} & 0 \\ 0 & 0 & 1 \end{bmatrix}$$

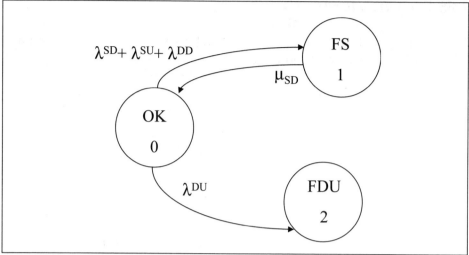

Figure F-18. 1oo1D Markov Model

The equation for 1oo1D MTTF is obtained in a manner similar to the 1oo1. After this process, it is discovered that this equation is identical to Equation B-4. This makes sense. The 1oo1D architecture merely converts dangerous failures to safe failures, it does not provide any fault tolerance.

$$MTTF = \frac{1}{\lambda^S + \lambda^D}$$

Example Calculations for the 1oo1D Architecture

In order to compare architectures accurately, the failure rate for the diagnostic channel must be estimated as it is likely that extra components will be required. These failure rates must be added to those for the single board PEC to obtain totals. Table F-8 shows the model parameters including the new failure rates.

Table F-8. Model Parameters for 1oo1D Architecture

Model Parameters				Total	
λ_{PEC}^{SD}	9996 FITS	λ_{DIAG}^{SD}	420 FITS	λ_{1oo1D}^{SD}	10416 FITS
λ_{PEC}^{SU}	98 FITS	λ_{DIAG}^{SU}	30 FITS	λ_{1oo1D}^{SU}	128 FITS
λ_{PEC}^{DD}	5956 FITS	λ_{DIAG}^{DD}	48 FITS	λ_{1oo1D}^{DD}	6004 FITS
λ_{PEC}^{DU}	112 FITS	λ_{DIAG}^{DU}	2 FITS	λ_{1oo1D}^{DU}	114 FITS
RT	8 Hours				
TI	8760 Hours				
SD	24 Hours				

When these numbers are substituted into the P matrix, the numerical matrix is:

$$P = \begin{pmatrix} & 0 & 1 & 2 \\ 0 & 0.99998334 & 0.00001655 & 0.00000011 \\ 1 & 0.04166667 & 0.95833333 & 0.00000000 \\ 2 & 0.00000000 & 0.00000000 & 1.00000000 \end{pmatrix}$$

A complete set of results for both the approximation method and Markov solutions are shown in Table F-9. When comparing the 1oo1D with the 1oo1 results of Table F-5, the PFD values are lower although not significantly. The PFS values are higher. The MTTF numbers are slightly lower indicating the effect of the extra diagnostic circuitry.

Table F-9. Results from 1oo1D Architecture Calculations

Calculation results for 1oo1D Architecture		
FT Equation	PFD	0.00099864
FT Equation	PFDavg	0.00049932
FT Equation	PFS	0.00039715
Markov Solution	PFD	0.00099775
Markov Solution	PFS	0.00039660
Markov Solution	MTTF	60,017 Hrs.

2oo3 Architecture

An architecture designed to tolerate both "safe" and "dangerous" failures is the 2oo3 (two units out of three are required for the system to operate). This architecture provides both safety and high availability with three controller units.

Two outputs from each controller unit are required for each output channel. The two outputs from the three controllers are wired in a "voting" circuit, which determines the actual output (Figure F-20). The output will equal the "majority." When two sets of outputs conduct, the load is energized. When two sets of outputs are off, the load is de-energized.

A closer examination of the voting circuit shows that it will tolerate a failure of either failure mode – dangerous (short circuit) or safe (open circuit). Figure F-21 shows that when one unit fails open circuit, the system effectively degrades to a 1oo2 configuration. If one unit fails short circuit the system effectively degrades to a 2oo2 configuration. In both cases, the system remains in successful operation.

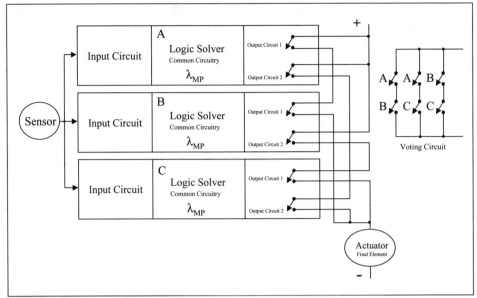

Figure F-19. 2oo3 Architecture

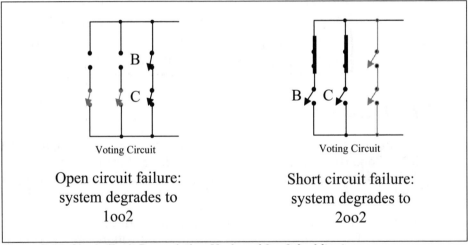

Figure F-20. Single Fault Degradation Modes of 2oo3 Architecture

PFD Fault Tree for 2oo3

The 2oo3 architecture will fail dangerously only if two units fail dangerously (Figure F-22). There are three ways in which this can happen, the AB leg can fail short circuit, the AC leg can fail short circuit and the BC leg can fail short circuit. These are shown in the top level events of the PFD fault tree of Figure F-23.

Each leg consists of two switches wired in series like a 1oo2 configuration. The subtree for each leg is developed for the 1oo2 configuration and each

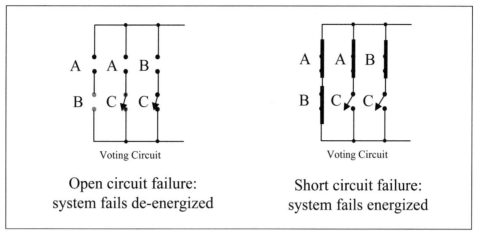

Figure F-21. Dual Fault Failure Modes of 2oo3 Architecture

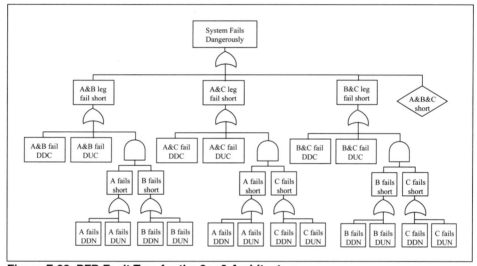

Figure F-22. PFD Fault Tree for the 2oo3 Architecture

looks like Figure F-12, the 1oo2 PFD fault tree. It should be noted that the system will also fail if all three legs fail dangerously. This can happen due to common cause or a combination of three independent failures. Since this is a second order effect, it can assumed to be negligible for first order approximation purposes. This is indicated in the fault tree with the "incomplete event" symbol. An approximation equation for PFD can be derived from the fault tree. The first order approximate equation for PFD is:

$$PFD_{2oo3} = 3\lambda^{DUC} \times TI + 3\lambda^{DDC} \times RT + 3(\lambda^{DDN} \times RT + \lambda^{DUN} \times TI)^2 \quad \text{(F-16)}$$

The equation for PFDavg is:

$$PFDavg_{2oo3} = 3\lambda^{DUC} \times TI/2 + 3\lambda^{DDC} \times RT + 3[(\lambda^{DDN} \times RT)^2$$
$$+ (\lambda^{DDN} \times RT \times \lambda^{DUN} \times TI)/2 + (\lambda^{DUN} \times TI)^2/3] \quad (F-17)$$

It should be noted that the simplified equations derived by approximation techniques are only valid for low failure rates.

PFS Fault Tree for 2oo3

The 2oo3 is a symmetrical architecture that successfully tolerates a short circuit or an open circuit failure. It will fail with outputs de-energized only when two failures occur as shown in Figure F-23. The fault tree for safe failures is shown in Figure F-24. It looks like the fault tree for dangerous failures except that the failure modes are different. This is the result of the symmetrical nature of the architecture. Note that each major event in the top level of the fault tree is equivalent to the 2oo2 fault tree of Figure F-20.

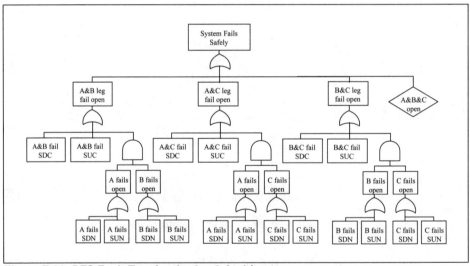

Figure F-23. PFS Fault Tree for the 2oo3 Architecture

The approximate equation for PFS derived from the fault tree (with detail added, Figure F-24) is:

$$PFS_{2oo3} = 3\lambda^{SDC} \times SD + 3\lambda^{SUC} \times SD + 6(\lambda^{SDN} \times RT \times \lambda^{SDN} \times SD$$
$$+ \lambda^{SUN} \times TI \times \lambda^{SDN} \times SD + \lambda^{SDN} \times RT \times \lambda^{SUN} \times SD + \lambda^{SUN} \times TI \times \lambda^{SUN} \times SD) \quad (F-18)$$

Appendix F: System Architectures   341

Markov Model for 2oo3

A Markov model can be created for the 2oo3 architecture. The model construction begins with all three units fully operational in state 0. All four normal failure modes of the three units must be placed in the model as exit rates from state 0. In addition, the four failure modes of common cause failures must also be included in the model. Since there are three combinations of two units, AB, AC and BC, three sets of common cause failures are included. (Much like the three sets in the fault tree diagrams.) When all these failure rates are placed, the partial model looks like Figure F-24. In state 1, one controller has failed in a safe detected manner. In state 2, one controller has failed in a safe undetected manner. In both states 1 and 2 the system has degraded to a 1oo2. In state 3, one controller has failed in a dangerous detected manner. In state 4, one controller has failed in a dangerous undetected manner. The system has degraded to a 2oo2 in states 3 and 4. The system has not failed in any of these states.

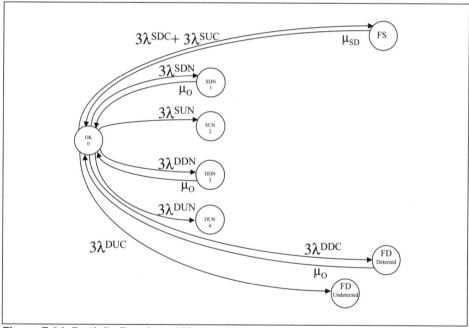

Figure F-24. Partially Developed Markov Model for the 2oo3 Architecture

In states 1 and 2 the system is operating in 1oo2 mode. Further safe failures will fail the system with outputs de-energized (safe). Further dangerous failures lead to secondary degradation. From states 3 and 4 the system is operating in 2oo2 mode. Additional dangerous failures fail the system with outputs energized. Further safe failures degrade the system again. Repair rates are added to the diagram. It is assumed that the system is inspected and that all failed units are repaired if a service call is made. Therefore, all repair rates transition to state 0.

Appendix F: System Architectures

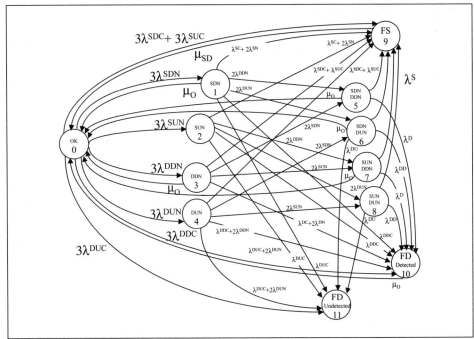

Figure F-25. 2oo3 Markov Model

The transition matrix for Figure F-25 is:

$$P = \begin{bmatrix}
1-\Sigma & 3\lambda^{SDN} & 3\lambda^{SUN} & 3\lambda^{DDN} & 3\lambda^{DUN} & 0 & 0 & 0 & 0 & 3\lambda^{SC} & 3\lambda^{DDC} & 3\lambda^{DUC} \\
\mu_O & 1-\Sigma & 0 & 0 & 0 & 2\lambda^{DDN} & 2\lambda^{DUN} & 0 & 0 & \lambda^{SC}+2\lambda^{SN} & \lambda^{DDC} & \lambda^{DUC} \\
0 & 0 & 1-\Sigma & 0 & 0 & 0 & 0 & 2\lambda^{DDN} & 2\lambda^{DUN} & \lambda^{SC}+2\lambda^{SN} & \lambda^{DDC} & \lambda^{DUC} \\
\mu_O & 0 & 0 & 1-\Sigma & 0 & 2\lambda^{SDN} & 0 & 2\lambda^{SUN} & 0 & \lambda^{SC} & \lambda^{DC}+2\lambda^{DN} & 0 \\
0 & 0 & 0 & 0 & 1-\Sigma & 0 & 2\lambda^{SDN} & 0 & 2\lambda^{SUN} & \lambda^{SC} & \lambda^{DDC}+2\lambda^{DDN} & \lambda^{DUC}+2\lambda^{DUN} \\
\mu_O & 0 & 0 & 0 & 0 & 1-\Sigma & 0 & 0 & 0 & \lambda^{S} & \lambda^{D} & 0 \\
\mu_O & 0 & 0 & 0 & 0 & 0 & 1-\Sigma & 0 & 0 & \lambda^{S} & \lambda^{DD} & \lambda^{DU} \\
\mu_O & 0 & 0 & 0 & 0 & 0 & 0 & 1-\Sigma & 0 & \lambda^{S} & \lambda^{D} & 0 \\
0 & 0 & 0 & 0 & 0 & 0 & 0 & 0 & 1-\Sigma & \lambda^{S} & \lambda^{DD} & \lambda^{DU} \\
\mu_{SD} & 0 & 0 & 0 & 0 & 0 & 0 & 0 & 0 & 1-\Sigma & 0 & 0 \\
\mu_O & 0 & 0 & 0 & 0 & 0 & 0 & 0 & 0 & 0 & 1-\Sigma & 0 \\
0 & 0 & 0 & 0 & 0 & 0 & 0 & 0 & 0 & 0 & 0 & 1
\end{bmatrix}$$

where:

$1-\Sigma$ indicates one minus the sum of all other row elements

Reliability and safety metrics can be calculated using matrix techniques from this P matrix.

Example Calculations for the 2oo3 Architecture

Substituting numerical values into the P matrix yields:

P	0	1	2	3	4	5	6	7	8	9	10	11
0	0.9999498	3.039E-05	4.057E-07	1.81E-05	3.41E-07	0	0	0	0	6.2844E-07	3.69E-07	6.96E-09
1	0.125	0.8749668	0	0	0	1.207E-05	2.274E-07	0	0	2.0739E-05	1.23E-07	2.32E-09
2	0	0	0.9999668	0	0	0	0	1.207E-05	2.274E-07	2.0739E-05	1.23E-07	2.32E-09
3	0.125	0	0	0.8749668	0	2.026E-05	0	2.705E-07	0	2.0948E-07	1.24E-05	0
4	0	0	0	0	0.9999668	0	2.026E-05	0	2.705E-07	2.0948E-07	1.22E-05	2.3E-07
5	0.125	0	0	0	0	0.8749833	0	0	0	1.0474E-05	6.27E-06	0
6	0.125	0	0	0	0	0	0.8749833	0	0	1.0474E-05	6.16E-06	1.16E-07
7	0.125	0	0	0	0	0	0	0.8749833	0	1.0474E-05	6.27E-06	0
8	0	0	0	0	0	0	0	0	0.9999833	1.0474E-05	6.16E-06	1.16E-07
9	0.0416667	0	0	0	0	0	0	0	0	0.95833333	0	0
10	0.125	0	0	0	0	0	0	0	0	0	0.875	0
11	0	0	0	0	0	0	0	0	0	0	0	1

Truncating the rows and columns of the failure states will provide the Q matrix. This matrix is subtracted from the identity matrix to obtain:

I-Q	0	1	2	3	4	5	6	7	8	9
0	5.024E-05	-3.039E-05	-4.057E-07	-1.81E-05	-3.41E-07	0	0	0	0	-6.284E-07
1	-0.125	0.1250332	0	0	0	-1.207E-05	-2.274E-07	0	0	-2.074E-05
2	0	0	3.316E-05	0	0	0	0	-1.207E-05	-2.274E-07	-2.074E-05
3	-0.125	0	0	0.1250332	0	-2.026E-05	0	-2.705E-07	0	-2.095E-07
4	0	0	0	0	3.316E-05	0	-2.026E-05	0	-2.705E-07	-2.095E-07
5	-0.125	0	0	0	0	0.1250167	0	0	0	-1.047E-05
6	-0.125	0	0	0	0	0	0.1250167	0	0	-1.047E-05
7	-0.125	0	0	0	0	0	0	0.1250167	0	-1.047E-05
8	0	0	0	0	0	0	0	0	1.675E-05	-1.047E-05
9	-0.0416667	0	0	0	0	0	0	0	0	0.04166667

The I - Q matrix is inverted to obtain the N matrix.

N	0	1	2	3	4	5	6	7	8	9
0	1962641.9	476.99707	24015.477	284.09384	20186.923	0.092073	3.2720931	2.318424	652.11499	42.0604415
1	1962640	484.99446	24015.453	284.09355	20186.902	0.0928448	3.2721043	2.3184216	652.11433	42.06438
2	1950140.5	473.95875	54021.977	282.28425	20058.338	0.0914865	3.2512509	5.2144419	1057.4356	56.9073007
3	1962447	476.94969	24013.091	292.0635	20184.917	0.0933599	3.2717681	2.318211	652.05021	42.0563042
4	1221502.9	296.87193	14946.677	176.81343	42723.344	0.0573041	6.9237199	1.4429334	892.99454	26.4527544
5	1962543.5	476.97314	24014.272	284.07959	20185.91	8.0909968	3.2719289	2.3183077	652.08227	42.0603421
6	1962543.5	476.97314	24014.272	284.07959	20185.91	0.0920684	11.270857	2.3183077	652.08227	42.0603421
7	1962543.5	476.97314	24014.272	284.07959	20185.91	0.0920684	3.2719289	10.317236	652.08227	42.0603421
8	1227559.5	298.34392	15020.787	177.69013	12626.169	0.0575882	2.0465725	1.4500879	60123.627	41.3183485
9	1962641.9	476.99707	24015.477	284.09384	20186.923	0.092073	3.2720931	2.318424	652.11499	66.0604415

The MTTF is the sum of the state 0 row. The PFS and PFD are calculated by multiplying the P matrix by a row matrix S starting with the system in state 0.

The results of the calculations for the 2oo3 architecture are shown in Table F-10.

Appendix F: System Architectures

Table F-10. Results from 2oo3 Architecture Calculation

Calculation results for 2oo3 Architecture		
FT Equation	PFD	0.00006719
FT Equation	PFDavg	0.00003515
FT Equation	PFS	0.00001583
Markov Solution	PFD	0.00006674
Markov Solution	PFS	0.00001665
Markov Solution	MTTF	2,008,305 Hrs

2oo2D Architecture

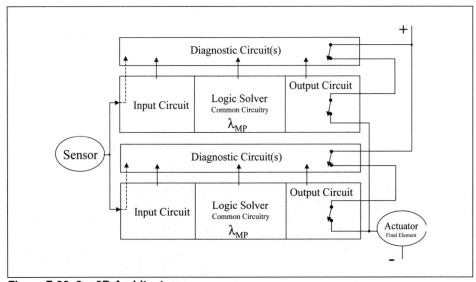

Figure F-26. 2oo2D Architecture

The 2oo2D is a four channel architecture that consists of two 1oo1D controllers arranged in a 2oo2 style (Figure F-26). Since the 1oo1D protects against dangerous failures when diagnostics detect the failure, two units can be wired in parallel to protect against shutdowns. Effective diagnostics are essential to this architecture as an undetected dangerous failure on either unit will fail the system dangerously.

PFD Fault Tree for 2oo2D

The 2oo2D architecture will fail with outputs energized if either unit has a dangerous undetected failure or if the system experiences a dangerous undetected common cause failure. This is shown in the fault tree of Figure F-27.

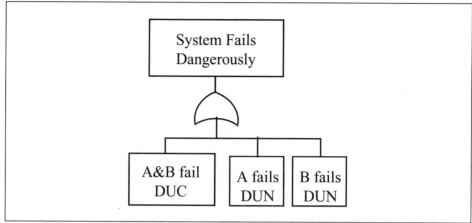

Figure F-27. PFS Fault Tree for the 2oo2D Architecture

The approximate equation derived from the fault tree for PFD is:

$$\text{PFD}_{2oo2D} = \lambda^{DUC} \times TI + 2\lambda^{DUN} \times TI \tag{F-19}$$

The equation for PFDavg is:

$$\text{PFDavg}_{2oo2D} = \lambda^{DUC} \times TI/2 + 2\lambda^{DUN} \times TI/2 \tag{F-20}$$

PFS Fault Tree for 2oo2D

Figure F-28 shows that a 2oo2D architecture will fail safely only if both units fail safely. This can happen due to common cause failures SDC, SUC, or DDC, or if A and B fail safely.

First order approximation techniques can be used to generate a formula for probability of failing safely from this fault tree.

$$\begin{aligned}\text{PFS}_{2oo2D} = &(\lambda^{SDC} + \lambda^{SUC} + \lambda^{DDC}) \times SD + 2((\lambda^{SDN} + \lambda^{DDN}) \times RT \\ &\times (\lambda^{SDN} + \lambda^{DDN}) \times SD + \lambda^{SUN} \times TI \times (\lambda^{SDN} + \lambda^{DDN}) \times SD \\ &+ (\lambda^{SDN} + \lambda^{DDN}) \times RT \times \lambda^{SUN} \times SD + \lambda^{SUN} \times TI \times \lambda^{SUN} \times SD)\end{aligned} \tag{F-21}$$

where

> SD = the time required to restart the process after a shutdown
> RT = average repair time for a detected failure
> TI = the inspection time interval

This equation has the same form as Equation B-10 with $(\lambda^{SDN} + \lambda^{DDN})$ substituted for λ^{SDN}. As in the case of the 2oo2 architecture, the equation development can be very confusing, as the fault tree does not show

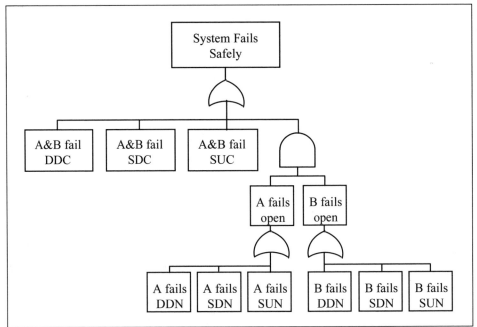

Figure F-28. PFS Fault Tree for the 2oo2D Architecture

sufficient detail. The fully developed fault tree is so complicated that it does not provide any insight into the operation of the architecture.

Markov Model for 2oo2D

The Markov model for a 2oo2D system is shown in Figure F-29.

Three system success states that are similar to the other dual systems previously developed are shown. State 1 is an interesting case. It represents a safe detected failure or a dangerous detected failure. The result of both failures is the same since the diagnostic cutoff switch de-energizes the output whenever a dangerous failure is detected. The only other system success state, state 2, represents the situation in which one controller has failed in a safe undetected manner. The system operates because the other controller manages the load.

The 2oo2D architecture shows good tolerance to both "safe" and "dangerous" failures. However, since coverage is utilized to convert dangerous failures into safe failures, this tolerance depends in great part

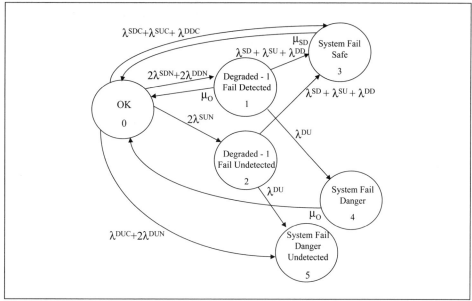

Figure F-29. 2oo2D Markov Model

on the diagnostic coverage. The transition matrix for the single-board Markov model of Figure F-30 is:

$$P = \begin{bmatrix} 1-\Sigma & (2\lambda^{SDN}+2\lambda^{DDN}) & 2\lambda^{SUN} & (\lambda^{SC}+\lambda^{DDC}) & 0 & (\lambda^{DUC}+2\lambda^{DUN}) \\ \mu_O & 1-\Sigma & 0 & (\lambda^{S}+\lambda^{DD}) & (\lambda^{DU}) & 0 \\ 0 & 0 & 1-\Sigma & (\lambda^{S}+\lambda^{DD}) & 0 & (\lambda^{DU}) \\ \mu_{SD} & 0 & 0 & 1-\mu_{SD} & 0 & 0 \\ \mu_O & 0 & 0 & 0 & 1-\mu_O & 0 \\ 0 & 0 & 0 & 0 & 0 & 1 \end{bmatrix}$$

Example Calculations for the 2oo2D Architecture

Using the four total failure rates from Table F-3, common cause must be added to the model because redundant units are present. Using the same Beta factor as before (Beta = 0.02), the failure rates are calculated in Table F-11.

Table F-11. Model Parameters Including Common Cause Failure Rates

Model Parameters	
λ_{1oo1D}^{SDC}	208.3 FITS
λ_{1oo1D}^{SUC}	2.6 FITS
λ_{1oo1D}^{SDN}	10207.7 FITS
λ_{1oo1D}^{SUN}	125.4 FITS
λ_{1oo1D}^{DDC}	120.1 FITS
λ_{1oo1D}^{DUC}	2.3 FITS
λ_{1oo1D}^{DDN}	5883.9 FITS
λ_{1oo1D}^{DUN}	111.7 FITS
Beta	0.02
RT	8 Hours
TI	8760 Hours
SD	24 Hours

Results can be obtained by substituting these values into the equations above using the approximation methods. The Markov model can be solved numerically as well. Substituting values into the P matrix results in:

P	0	1	2	3	4	5
0	0.99996701	0.00003218	0.00000025	0.00000033	0.00000000	0.00000023
1	0.12500000	0.87498334	0.00000000	0.00001655	0.00000011	0.00000000
2	0.00000000	0.00000000	0.99998334	0.00001655	0.00000000	0.00000011
3	0.04166667	0.00000000	0.00000000	0.95833333	0.00000000	0.00000000
4	0.12500000	0.00000000	0.00000000	0.00000000	0.87500000	0.00000000
5	0.00000000	0.00000000	0.00000000	0.00000000	0.00000000	1.00000000

This matrix is truncated and subtracted from the identity matrix to obtain:

I-Q		
0.00003299	-0.00003218	-0.00000025
-0.12500000	0.12501666	0.00000000
0.00000000	0.00000000	0.00001666

The I-Q matrix is inverted to obtain the N matrix.

N		
1231755.67	317.092444	18546.5647
1231591.5	325.049117	18544.0929
0	0	60016.8047

The MTTF value is the sum of the state 0 row.

The results from the 2oo2D calculations are shown in Table F-12.

Table F-12. Results from the 2oo2D Architecture Calculations

Calculation results for 2oo2D Architecture		
FT Equation	PFD	0.00197731
FT Equation	PFDavg	0.00098865
FT Equation	PFS	0.00000889
Markov Solution	PFD	0.00197381
Markov Solution	PFS	0.00000882
Markov Solution	MTTF	1,250,619 Hrs.

1oo2D Architecture

The 1oo2D architecture is similar to the 2oo2D architecture except that additional control lines are added to allow one unit to de-energize the other unit. A 1oo2D architecture is shown in Figure F-30. Note that this architecture is defined as normally operating with all four switches closed. The specific PEC implementation discussed earlier in this thesis is a variation of the 1oo2D and operates with only one set of switches closed at a time. This model does not assume comparison diagnostics are present.

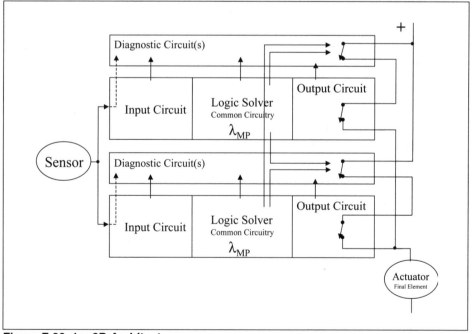

Figure F-30. 1oo2D Architecture

The primary difference in the 2oo2D versus the 1oo2D occurs when there is a dangerous undetected failure in one unit. Because of the added control lines and readback diagnostics, the operating unit can de-energize the failed unit. The 1oo2D architecture provides 1oo2 type functionality in this situation.

PFD Fault Tree for 1oo2D

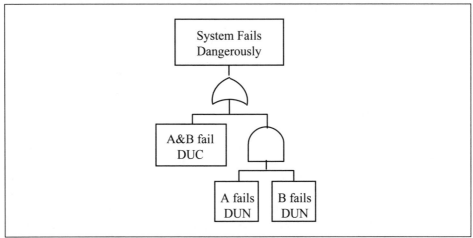

Figure F-31. PFD Fault Tree for the 1oo2D Architecture

The 1oo2D fails dangerously only if both units fail dangerously and that failure is not detected by the diagnostics in either unit. The fault tree is shown in Figure F-31. An approximate PFD equation developed from the fault tree is

$$\text{PFD}_{1oo2D} = \lambda^{DUC} \times TI + (\lambda^{DUN} \times TI)^2 \qquad \text{(F-22)}$$

This should be compared to Equation B-5 for the 1oo2 architecture. The 1oo2D provides better safety performance than the 1oo2 because only undetected failures are included in the PFD.

The equation for PFDavg for the 1oo2D architecture is derived from Equation B-20.

$$\text{PFDavg}_{1oo2D} = \lambda^{DUC} \times TI/2 + (\lambda^{DUN} \times TI)^2/3 \qquad \text{(F-23)}$$

PFS Fault Tree for 1oo2D

Figure F-32 shows that a 1oo2D architecture will fail safely if there is a common cause safe failure, a common cause dangerous detected failure, if both units fail in a detected manner or if both units fail in a safe undetected mode.

The approximation techniques can be used to generate a formula for probability of failing safely from this fault tree:

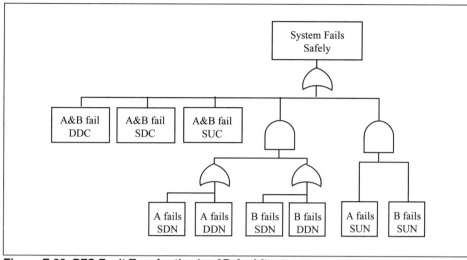

Figure F-32. PFS Fault Tree for the 1oo2D Architecture

$$PFS_{1oo2D} = (\lambda^{SDC} + \lambda^{SUC} + \lambda^{DDC}) \times SD$$
$$+ (\lambda^{SDN} \times RT + \lambda^{DDN} \times RT)^2 + (\lambda^{SUN} \times TI)^2 \qquad (F\text{-}24)$$

where

> SD = the time required to restart the process after a shutdown
> RT = average repair time for a detected failure
> TI = the inspection time interval

Markov Model for 1oo2D

The Markov model for a 1oo2D system is shown in Figure F-33. This model is developed for the situation where all four switches are closed in normal operation.

Four system success states are shown: 0, 1, 2 and 3. State 1 represents a safe detected failure or a dangerous detected failure. Like the 2oo2D, the result of both failures is the same since the diagnostic switch de-energizes the output whenever a dangerous failure is detected. Another system success state, state 2, represents the situation in which one controller has failed in a dangerous undetected manner. The system will operate correctly given a process demand because the other unit will detect the failure and de-energize the load via its 1oo2 control lines. The third system success state is shown in Figure F-34. One unit has failed with its output de-energized. The system load is maintained by the other unit that will still respond properly to a process demand.

In state 1 the system has degraded to 1oo1D operation. A second safe failure or a dangerous detected failure will fail the system safely. Like the 1oo1D, a dangerous undetected failure will fail the system dangerously. In

352   Appendix F: System Architectures

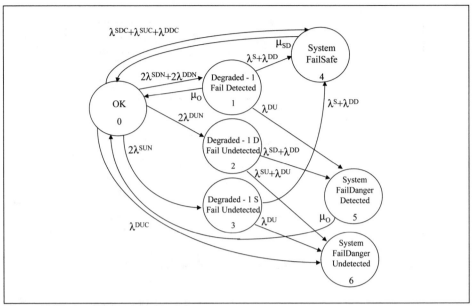

Figure F-33. 1oo2D Markov Model

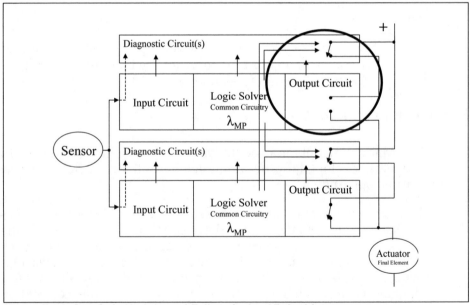

Figure F-34. One Safe Undetected Failure in 1oo2D Architecture

this situation though, the system fails to state 5 where one unit has failed detected and one unit has failed undetected. An assumption is that all units will be inspected and tested during a service call so an on-line repair rate exits from state 5 to state 0.

One unit has failed with a dangerous undetected failure in state 2. The system is still successful since it will respond to a demand as described above. From this state, any other component failure will fail the system dangerously. Since it is assumed that any component failure in a unit causes the entire unit to fail, it must be assumed that the control line from the lower unit to the upper unit will not work. Under those circumstances, the system will not respond to a process demand and is considered failed dangerously. Since the lower unit failed in a safe detected manner, the Markov model goes to state 5.

In state 3 one unit has failed in a safe undetected manner. In this condition the system has also degraded to 1oo1D operation. Additional safe failures or dangerous detected failures will cause the system to fail safely. An additional dangerous undetected failure will fail the system dangerously taking the Markov model to state 6 where both units have an undetected failure. Failures from this state are not detected until there is a maintenance inspection.

The P matrix for the Markov model of Figure F-32 is:

$$P = \begin{bmatrix} 1-\Sigma & 2\lambda^{SDN}+2\lambda^{DDN} & 2\lambda^{DUN} & 2\lambda^{SUN} & \lambda^{SC}+\lambda^{DDC} & 0 & \lambda^{DUC} \\ \mu_O & 1-\Sigma & 0 & 0 & \lambda^{S}+\lambda^{DD} & \lambda^{DU} & 0 \\ 0 & 0 & 1-\Sigma & 0 & 0 & \lambda^{SD}+\lambda^{DD} & \lambda^{SU}+\lambda^{DU} \\ 0 & 0 & 0 & 1-\Sigma & \lambda^{S}+\lambda^{DD} & 0 & \lambda^{DU} \\ \mu_{SD} & 0 & 0 & 0 & 1-\Sigma & 0 & 0 \\ \mu_O & 0 & 0 & 0 & 0 & 1-\Sigma & 0 \\ 0 & 0 & 0 & 0 & 0 & 0 & 1 \end{bmatrix}$$

where Σ represents the sum of the remaining row elements.

Example Calculations for the 1oo2D Architecture

Substituting numerical values into the P matrix yields:

P	0	1	2	3	4	5	6
0	0.999967009	0.000032183	0.000000223	0.000000251	0.000000331	0.000000000	0.000000002
1	0.125000000	0.874983338	0.000000000	0.000000000	0.000016548	0.000000114	0.000000000
2	0.000000000	0.000000000	0.999983338	0.000000000	0.000000000	0.000016420	0.000000242
3	0.000000000	0.000000000	0.000000000	0.999983338	0.000016548	0.000000000	0.000000114
4	0.041666667	0.000000000	0.000000000	0.000000000	0.958333333	0.000000000	0.000000000
5	0.125000000	0.000000000	0.000000000	0.000000000	0.000000000	0.875000000	0.000000000
6	0.000000000	0.000000000	0.000000000	0.000000000	0.000000000	0.000000000	1.000000000

Truncating and subtracting provides the numerical I - Q matrix.

I - Q				
	0.000032991	-0.000032183	-0.000000223	-0.000000251
1	-0.12500000	0.12501666	0.00000000	0.00000000
2	0.00000000	0.00000000	0.00001666	0.00000000
3	0.00000000	0.00000000	0.00000000	0.00001666

The N matrix is obtained by inverting the I - Q matrix.

$$N \quad \begin{matrix} 1231755.666 & 317.0924445 & 16518.03421 & 18546.56473 \\ 1231591.5 & 325.0491167 & 16515.83272 & 18544.09288 \\ 0 & 0 & 60016.80471 & 0 \\ 0 & 0 & 0 & 60016.80471 \end{matrix}$$

The calculation results for the 1oo2D architecture are shown in Table F-13.

Table F-13. Results from 1oo2D Calculations

Calculation results for 1oo2D Architecture		
FT Equation	PFD	0.00002093
FT Equation	PFDavg	0.00001031
FT Equation	PFS	0.00000917
Markov Solution	PFD	0.00002318
Markov Solution	PFS	0.00000882
Markov Solution	MTTF	1,267,137 Hrs.

Comparing Architectures

When the results of the analysis are examined, several of the main features of different architectures become apparent. Table F-14 compares the results. The availability parameter was added to the results. It is calculated with the equation

$$\text{Availability} = 1 - (\text{PFS} + \text{PFD})$$

It is interesting to note that the fault tree results and the Markov results are similar for this set of parameters. The failure rates used in the examples are sufficiently small to allow the first order approximation to be reasonably accurate.

Table F-14. Architecture Calculation Results

		1oo1	1oo2	2oo2	1oo1D	2oo3	2oo2D	1oo2D
FT Equation	PFD	0.00102877	0.00002159	0.00203696	0.00099864	0.00006719	0.00197731	0.00002093
FT Equation	PFDavg	0.00053821	0.00001110	0.00106565	0.00049932	0.00003515	0.00098865	0.00001031
FT Equation	PFS	0.00024226	0.00047967	0.00000528	0.00039715	0.00001583	0.00000889	0.00000917
FT Equation	Availability	0.99872898	0.99949874	0.99795776	0.99860421	0.99991697	0.99801380	0.99996990
Markov Solution	PFD	0.00102794	0.00002153	0.00203354	0.00099775	0.00006674	0.00197381	0.00002318
Markov Solution	PFS	0.00024195	0.00047898	0.00000524	0.00039660	0.00001665	0.00000882	0.00000882
Markov Solution	Availability	0.99873011	0.99949949	0.99796122	0.99860565	0.99991661	0.99801737	0.99996800
Markov Solution	MTTF	61874	49865	81543	60017	2008305	1250619	1267137

Looking at the Markov solutions, the 1oo2 architecture has the highest risk reduction factor (lowest PFD). The 2oo3 architecture has highest MTTF but the 1oo2D architecture has the highest availability. Both of these metrics indicate successful operation but show different views. The metrics indicate that the 2oo3 will have fewer system failures but will spend more time in a failed state.

REFERENCES AND BIBLIOGRAPHY

1. Goble, W.M. *Control Systems Safety Evaluation and Reliability*, Second Edition. ISA, 1998.

Appendix G
Modeling the Repair Process

Probability of Repair

Modeling the repair process can be complex and error prone. Fortunately, several assumptions can be made which simplify the analysis without serious impact on the results. However, one must understand the limits of these assumptions and simplifications.

In many ways modeling the repair process is difficult because the repair process is quite different from the failure process. Random failures are due to a stochastic process and most of our modeling techniques were created for these stochastic processes. Certain aspects of the repair process are deterministic. Other aspects of the repair process are stochastic. Fortunately, we can approximate the repair process more accurately with Markov models than most other techniques.

An estimate of repair probability can be made. Consider the single component Markov model of Figure G-1. Assume that failures are immediately detected when they occur. One can accurately model the repair process with a discrete time Markov model. Using a delta t of one hour, one must estimate the probability of repair for each hour after reaching state 1.

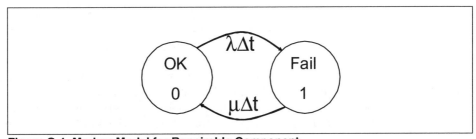

Figure G-1. Markov Model for Repairable Component

Assume that it is estimated that the repair probability will vary with time (non-homogeneous). Table G-1 shows a set of example repair time statistics. These statistics indicate that the repair time of a set of 64 repairs varied from one to six hours. Six units were repaired within one hour. Sixteen units were repaired within two hours. Other units took longer. Based on these numbers the average repair time is approximately 3 hours and the probability of repair in a particular hour is shown in Figure G-2.

Table G-1. Repair Time Statistics

Repair Hours	Quantity	Probability	Cumulative Probability	Total Repair Time
1	6	0.09	0.09	6
2	16	0.25	0.34	32
3	22	0.34	0.69	66
4	14	0.22	0.91	56
5	4	0.06	0.97	20
6	2	0.03	1.00	12
Total	64			192
	Average Repair Time			3

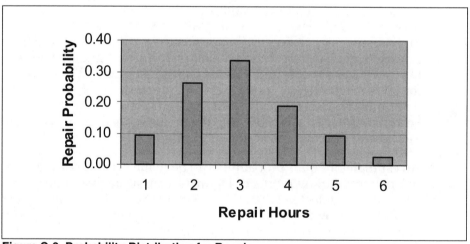

Figure G-2. Probability Distribution for Repair

Using an average repair rate of 0.333 (1/3 hours), a simple steady state availability equation (Figure G-3) gives an answer of 0.9708. Most would agree that the simple approximation is good enough.

The effect of the approximation can be shown clearly by creating another Markov model that provides exactly three hours of repair time. This model is shown in Figure G-4. The probability of successful operation (probability of being in state 0) for this model is 0.9708, the exact same value as the simple model.

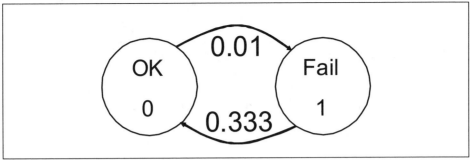

Figure G-3. Simple Approximate Markov Model

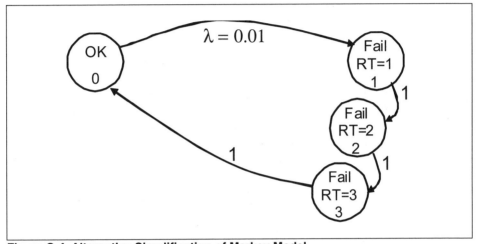

Figure G-4. Alternative Simplification of Markov Model

One Repairman, Two Repairman

When multiple components are modeled in a system, the issue of modeling multiple repairs arises. Consider a dual unit modeled by the simple, one failure mode Markov model of Figure G-5.

The probability of moving from state 2 to state 1 has been modeled two different ways from various reasons. If the constant repair rate for one unit is Mu (μ), then the probability of moving from state 1 to state 0 is equal to Mu given a delta t of one hour. The probability of moving from state 2 to state 1 has been modeled as either 2 Mu (often called the "two repairman model") or Mu (often called the "one repairman model").

One argument for using a value of 2 Mu assumes that two independent repair crews are available (see Ref. 1). Therefore, if the failures are due to a random, stochastic processes, then repairs can occur at the same time increasing the repair rate to 2 Mu. The argument in favor of using the value of Mu is that typically only one repair crew is available, and therefore it is more conservative to use the single Mu value.

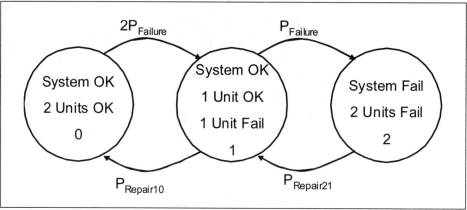

Figure G-5. Two Component Markov Model

An alternative argument (see Ref. 2) has been made for using the value of 2 Mu. This argument recognizes that only a single repair crew is available. However, the second failure occurs only from state 1 where one component is already failed and is likely under repair. Since the repair of the first component is on average half complete, that repair will take only half the time to complete. When that repair is complete the Markov model is back in state 1. Therefore, the repair probability from state 2 is twice as likely than the repair rate from state 1 to state 0. This is a reasonable argument and justifies the use of 2 Mu in situations where all failures are immediately detectable.

Periodic Inspection, Test and Repair

In situations where a failure is not detectable because of automatic diagnostics or an overt indication, the modeling of repair must be done quite differently. It must be remembered that the failure must first be detected then repaired. Often in Safety Instrumented System applications, detection of some failure types are done when a periodic inspection and test is performed.

Many use an unreliability technique to model this situation since the "mission time" is equivalent to the time period between inspections. The probability of failure is set to zero after the inspection, test and repair if it is assumed that the inspection and repair process is perfect. That results in probability of failure plots that look like Figure G-6.

This situation can be modeled using probability combination techniques or with Markov models. Consider the multiple failure state Markov model of Figure G-7. This model shows a simplified 1oo2 system without common cause or diagnostics. (For a complete model of a 1oo2 system, see Appendix F.)

Appendix G: Modeling the Repair Process   361

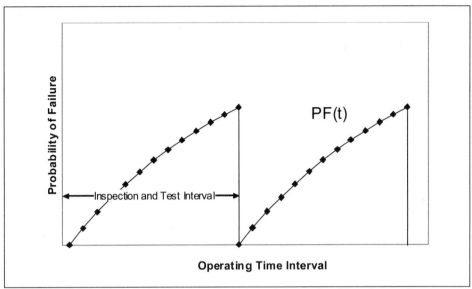

Figure G-6. Probability of Failure as a Function of Operating Time Interval

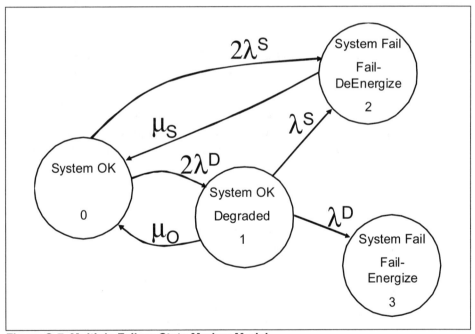

Figure G-7. Multiple Failure State Markov Model

This model can be solved using discrete time matrix multiplication for the case where periodic inspection and test is done to detect failures in state 3. The P matrix is normally:

$$\begin{bmatrix} 1-(2\lambda^D+2\lambda^S) & 2\lambda^D & 2\lambda^S & 0 \\ \mu_O & 1-(\mu_O+\lambda^S+\lambda^D) & \lambda^S & \lambda^D \\ \mu_S & 0 & 0 & 0 \\ 0 & 0 & 0 & 1 \end{bmatrix}$$

When the time counter equals the end of the inspection, test and repair period the matrix is changed to represent the known probabilities of failure. The P matrix (PTI matrix) used then is:

$$\begin{bmatrix} 1 & 0 & 0 & 0 \\ 1 & 0 & 0 & 0 \\ 1 & 0 & 0 & 0 \\ 1 & 0 & 0 & 0 \end{bmatrix}$$

This matrix represents the probability of detecting and repairing the failure. The use of "1" indicates the assumption of perfect inspection and repair.

When this approach is used the plot of probability as a function of time is shown in Figure G-8. Note that an assumption has been made that the system provides functionality during the inspection and testing. That assumption may not be realistic.

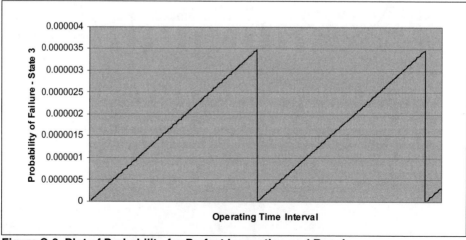

Figure G-8. Plot of Probability for Perfect Inspection and Repair

Modeling Imperfect Inspection, Test and Repair

In the previous analysis two assumptions were made that are not realistic. Both can have a significant impact on results. The assumption of perfect inspection, test and repair can easily be changed when solving the models using discrete time matrix multiplication with a Markov model. To include the effect of imperfect inspection and repair, the failure rate going to the fail-energize (absorbing) state can be split into those detected during a periodic inspection and those that are not. The upgraded Markov model is shown in Figure G-9.

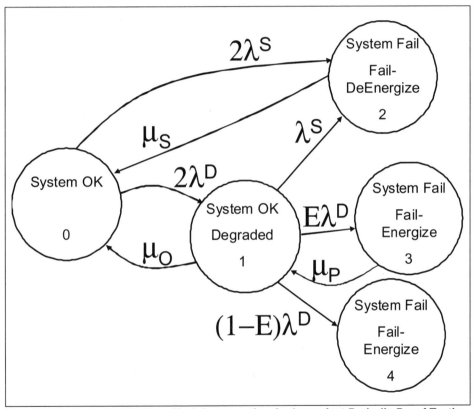

Figure G-9. Upgraded Markov Model accounting for Imperfect Periodic Proof Testing

The P matrix is normally:

$$\begin{bmatrix} 1-(2\lambda^D+2\lambda^S) & 2\lambda^D & 2\lambda^S & 0 & 0 \\ \mu_O & 1-(\mu_O+\lambda^S+\lambda^D) & \lambda^S & E\lambda^D & (1-E)\lambda^D \\ \mu_S & 0 & 0 & 0 & 0 \\ 0 & 0 & 0 & 1 & 0 \\ 0 & 0 & 0 & 0 & 1 \end{bmatrix}$$

When the time counter equals the end of the inspection, test and repair period the matrix is changed to represent the known probabilities of failure. The P matrix (PTI matrix) used then is:

$$\begin{bmatrix} 1 & 0 & 0 & 0 & 0 \\ 1 & 0 & 0 & 0 & 0 \\ 1 & 0 & 0 & 0 & 0 \\ 1 & 0 & 0 & 0 & 0 \\ 0 & 0 & 0 & 0 & 1 \end{bmatrix}$$

This matrix indicates that all failures are detected and repaired except those from state 4 where they remain failed. A plot of the PFD as a function of operating time interval is shown in Figure G-10.

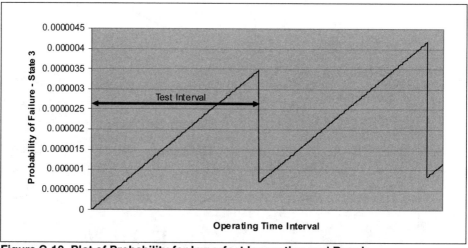

Figure G-10. Plot of Probability for Imperfect Inspection and Repair

Modeling the Time Period for Inspection, Test and Repair Periods

When a periodic inspection, test and repair procedure takes a significant amount of time, the time period may have an effect on the probability of failure. This is not an issue when the safety instrumented system remains on-line and capable of responding to a demand. But if the system must be placed in bypass for the time period when the test is run, that may be important. Fortunately, this can also easily be modeled with a discrete time matrix Markov model. At the beginning of the inspection, test and repair period an alternative P matrix is used which assigns the probability of failure in state 3 to "1." This represents the situation where the system is disabled during the test. The matrix used during the test period would be:

$$\begin{bmatrix} 0 & 0 & 0 & 1 \\ 0 & 0 & 0 & 1 \\ 0 & 0 & 0 & 1 \\ 0 & 0 & 0 & 1 \\ 0 & 0 & 0 & 1 \end{bmatrix}$$

A plot of probability of failure in state 3 is shown in Figure G-11.

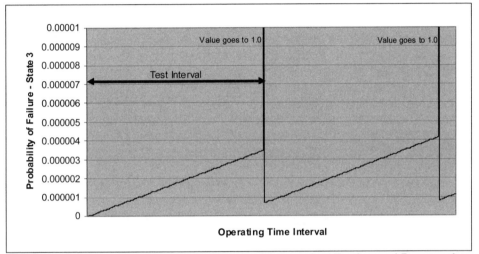

Figure G-11. Probability of Failure in State 3 with Imperfect Testing and Bypassed Testing

Overall, more sophisticated models could be used but the effect of repair is well approximated when the modeling includes:

- Proof test interval
- Proof test coverage
- Proof test time – bypass or no bypass
- Actual repair time

These variables are often not included in simplified equations and this can clearly result in probabilistic verification calculations that are optimistic leading to insufficient safety.

REFERENCES AND BIBLIOGRAPHY

1. Bukowski, J. V. "A Comparison of Techniques for Computing PFD Average." *2005 Proceedings of the Annual Reliability and Maintainability Symposium.* IEEE, 2005.

2. Bukowski, J. V. Notes during private meeting with author. September 2004.

Appendix H
Answers to Exercises

Chapter 1: Safety Life Cycle

1-1. b. false. The standards only require that the general requirements of the safety lifecycle as outlined in the respective standard be followed.

1-2. e. All of the above

1-3. d. Satisfying legal requirements, meeting customer demands and achieving cost savings are all likely outcomes of following the IEC 61511 or IEC61508 standard. The challenge in realizing the cost savings is in how a firm chooses to use the standard most efficiently and effectively for safety system design and use.

1-4. d. Process hazards analysis is step 3 of the IEC safety lifecycle falling just after scope definition and just before overall specifying overall system requirements.

1-5. b.

1-6. b. SIL Verification Analysis is not usually part of the analysis phase. SIL verification is conducted after a safety instrumented function conceptual design has been put forward. This step should only occur after the safety requirements specification is complete at the end of the analysis phase of the SLC.

1-7. d. Functional safety management applies to all phases of the safety lifecycle.

1-8. b. Analyze Risk

1-9. c. The SLC ends only when the safety instrumented system is decommissioned and taken out of service.

Chapter 2: Safety Instrumented Systems

2-1. a. The physical design of the system and the ability to be located in an environment with flammable material are important characteristics for both control systems and safety systems.

2-2. c.

2-3. d. all of the above

2-4. d. fail-safe

2-5. a. A SIS most typically acts to reduce the likelihood of a harmful event by sensing a potentially unsafe condition and acting to bring the system to a safe state. It is also possible to use a SIS to reduce the magnitude of harm although this is less typical.

2-6. b. Both of the IEC standards and the ISA standard specifically cover SIL selection methods. ISO 9000 addresses quality work process and production system issues.

2-7. a. Yes – it reduces risk

2-8. The answer will vary depending on the definition of the SIF. One must know the hazard to be sure which equipment is part of the safety function. If it is assumed that pump protection is not the hazard and process disturbance is not the hazard, then Valve 2 and the pump motor contactors are not included.

Chapter 3: Equipment Failure

3-1. Software bugs are considered systematic failures.

3-2. $1 - (0.85)^{10} = 0.803$

3-3. $1 - 0.002 = 0.998$

3-4. $\lambda = 1/75 = 0.0133$ failures / year

3-5. $R = e^{(-0.0133 \times 0.5)} = 0.9933$

3-6. MTTF $= 1/0.006 = 166.6$ years

3-7. The correct answer is d. All of the common pieces of equipment are best analyzed by statistics while more unusual equipment is best analyzed by fault propagation type models.

3-8. The correct answer is e, none of the above.

3-9. $\lambda = 16 / ((50-16) \times 7640) = 6.16\text{E-}5$ /hour. This assumes the worst case assumption that all 16 units fail in the first hour.

3-10. To obtain a worst case failure rate, one could assume that each unit will operate for one additional hour. The MTTF is that case if 7641 hours. The failure rate would be 1 / 7641 failures per hour.

Chapter 4: Basic Reliability Engineering

4-1. $U = 1 - e^{(-0.015 \times 5)} = 0.072$

U(approximation) = $0.015 \times 5 = 0.075$

4-2. Lambda = 0.015 f/y

MTTF = 66.6 years
MTTR = 24 hours
MTTR = 0.00274 years
SS Unav. = 1- (MTTF/(MTTF + MTTR)) = 4.10942E-05

4-3. Lambda = 0.015 f/y

TI = 1 year
PFavg = $0.015 \times 1/2 = 0.0075$ (perfect inspection)

4-4. Lambda = 0.06 f/y

TI = 1 year
Cpt = 0.6
LT = 10 years
PFavg = $0.6 \times 0.06 \times 1/2 + 0.4 \times 0.06 \times 10/2 = 0.138$

4-5. b. An estimate of probability can be made by dividing the number of failures by the number of trails. Note: There is a level of uncertainty in this number due to the low number of trails. This can be calculated but is outside the scope of this chapter.

4-6. a.

Probability of success of transmitter = 1 − 0.15 = 0.85
Probability of success of controller = 1 − 0.008 = 0.992
Probability of success of valve = 1 − 0.19 = 0.81

System success requires successful operation of all three components. Therefore,

Probability of system success = $0.85 \times 0.992 \times 0.81 = 0.6829$

4-7. MTBF = MTTF + MTTR = 28,048 hours

4-8.   PFDavg = 5E-6 × 8760 × 2 / 2 = 0.0438

4-9.   The best measure to use is steady state unavailability.

4-10. The best measure to use is the average probability of failure. This should be calculated using a periodic repair function.

Chapter 5: System Reliability Engineering

5-1.   Solution

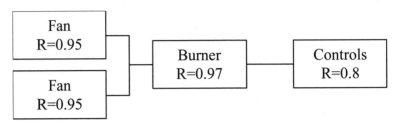

R Fan sub-system = RFan1 + RFan2 − (RFan1 × RFan2)
            = 0.9 + 0.9 − (0.9 × 0.9) = 0.99
R System = R Fan sub-system × R Burner × R Controls
RSystem = 0.99 × 0.98 × 0.85 = 0.8247

5-2.   a. Add the failure rates, total = 0.131

5-3.   e. R = $e^{(-0.131 \times 1)}$ = 0.877

5-4.   b. A = 0.99 × 0.98 = 0.9702

5-5.   b. 0.99646

5-6. Solution

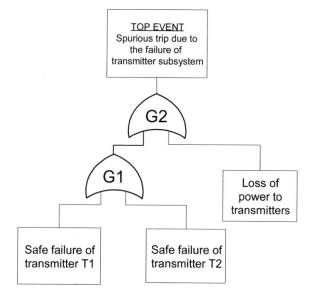

$G1 = (0.02 + 0.02) - (0.02 \times 0.02) = 0.0396/\text{year}$
$G2 = (0.0396 + 0.01) - (0.0393 \times 0.01) = 0.0492/\text{year}$

5-7.

$$
\begin{array}{c}
\text{From:} \\
\begin{array}{c} 0 \\ 1 \\ 2 \\ 3 \\ 4 \\ 5 \end{array}
\end{array}
\overset{\text{To:} \quad 0 \qquad\quad 1 \qquad\qquad 2 \qquad\quad 3 \qquad\qquad 4 \qquad\quad 5}{
\begin{bmatrix}
1-(\lambda_1+\lambda_2+\lambda_3+\lambda_4) & \lambda_2 & \lambda_3 & \lambda_1 & 0 & \lambda_4 \\
\mu_1 & 1-(\mu_1+\lambda_5+\lambda_6) & 0 & \lambda_5 & \lambda_6 & 0 \\
\mu_2 & 0 & 1-(\mu_2+\lambda_7) & 0 & 0 & \lambda_7 \\
0 & 0 & 0 & 1 & 0 & 0 \\
0 & \mu_3 & 0 & \lambda_8 & 1-(\mu_3+\lambda_8+\lambda_9) & \lambda_9 \\
0 & 0 & 0 & 0 & 0 & 1
\end{bmatrix}}
$$

5-8.

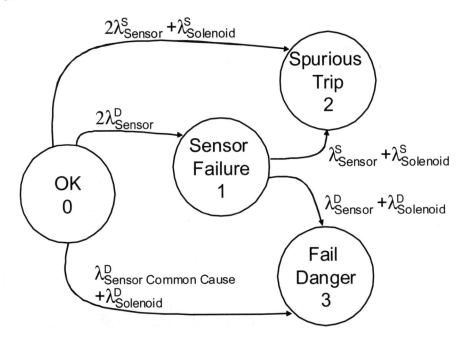

Chapter 6: Equipment Failure Modes

6-1. This failure would be classified as fail safe

6-2. PFD = $(\lambda_{dd} \times RT) + (\lambda_{du} \times T) = 0.00104$
PFDavg = $(\lambda_{dd} \times RT) + (\lambda_{du} \times T/2) = 0.00054$

6-3. $\lambda_D = (1-\%Safe) \times \lambda = 0.5 \times 1.2E-6 = 0.6E-6$

6-4. This failure would be classified as dangerous

6-5. This failure would be classified as dangerous

6-6. This failure would be classified as Fail safe

6-7. This failure would be classified as "No effect"

6-8. This failure would be classified as dangerous

Chapter 7: SIL Verification

7-1. This SIF be classified as low demand as two five year inspection periods occur during the expected ten year demand interval.

7-2. The demand mode is high demand as demands are expected every minute.

Although automatic diagnostics occur twice a minute, one cannot be sure from the information given that total diagnostic and response time will provide protection when failures are detected by the diagnostics. The total dangerous failure rate would be 9.0E-6 failures per hour. That meets the requirements for SIL1.

7-3. High demand. Dangerous detected failures will be converted into safe failures with a good margin on timing. The remaining dangerous failure rate is 0.5E-6 failures per hour. That meets the requirements for SIL2.

7-4. $\lambda_D = (1-\%Safe) \times \lambda$
$= 0.6 \times 6E-6$
$= 3.6E-6$ failures/hr

7-5. $$SFF = \frac{\lambda^{SD} + \lambda^{SU} + \lambda^{DD}}{\lambda^{SD} + \lambda^{SU} + \lambda^{DD} + \lambda^{DU}}$$
$= 97.9\%$

For a Type B instrument with a SFF of 97.9%, a 1oo1 architecture would qualify for SIL 2.

7-6. HFT = 1

7-7. HFT = 1

7-8. As per IEC 61508 - HFT = 1, Type B, SIL level = 3

As per ANSI/ISA 84.00.01-2004 (IEC 61511 Mod.) without sufficient "prior use" evidence, SIL level = 2. With sufficient "prior use" evidence SIL level = 3.

Chapter 8: Getting Failure Rate Data

8-1. **FMEA** stands for Failure Modes and Effects Analysis. It is a systematic procedure designed to find design issues by looking at all failure modes of components and reviewing the effects of the failures.

FMEDA stands for Failure Modes, Effects and Diagnostic Analysis. It is an extension of FMEA Technique by reviewing if the failure mode(s) are detectable and calculating the associated frequencies.

8-2. The main reasons are:

- Field data may include both systematic and random failures

- Field data may include failures occurring during the wearout portion of the bathtub curve

8-3. No, more data is usually required to carry out SIL verification calculations, e.g. SFF, dangerous detected failure rates, etc. In addition, the methods used to calculate the failure rate are missing. The given number is very suspicious and seems quite optimistic.

8-4. Failure/hr = 15.6×10^{-9} = 1.56E-8 f/hr

8-5. All of the solenoid.

8-6. Some limitations are:

- Specific field and environmental conditions are missed
- Systematic failures not addressed
- Installation problems may be missed

8-7. No

8-8. 600 FITS

Chapter 9: SIS Sensors

9-1. a.

9-2. c.

9-3. Detectors used to detect ultraviolet emissions from the flame are normally used for BMS. These detectors are specially designed to detect loss of flame.

9-4. The criterion is based on needed risk reduction. If the required risk reduction is greater than 10, the function should be classified as a SIF.

9-5. The criterion is based on needed risk reduction. If the required risk reduction is greater than 10, the function should be classified as a SIF.

Chapter 10: Logic Solvers

10-1. Yes, the reliability of relays are acceptable for SIL 3 applications.

Some issues are:

- Use safety control relays. These relays are more reliable, provide better protection against contact welding, and are tamper proof.

- Should not be used for large systems
- Have limited or no diagnostics
- No communications

10-2. The "D" in the 2oo2D and 1oo1D denotes that there is a switch incorporated in the system controlled by diagnostics to improve the SIL and availability of the system. The main difference between a 2oo2D and a 1oo2D logic solver is when there is a dangerous undetected failure in a unit. In the 1oo2D, the good unit can de-energize the failed unit. This allows both the safety of the overall system to be improved at a cost in availability.

10-3. **Redundancy** is the use of multiple elements or systems to perform the same function.

Voting is related to the manner in with the redundant elements operate to increase the safety and/or reliability of the overall system.

An **Architecture** describes a system from a redundancy and voting point of view e.g. XooY refers to the an architecture.

10-4. The key differences relate to the manner in which the system will function when a single or multiple processor failures occurs e.g.

For 3-2-0 mode of operation is: 2oo3→1oo2 → shutdown
For 3-2-1-0 mode of operation is: 2oo3→1oo2 → 1oo1 → shutdown

10-5. There are requirements in IEC 61511 (Part 1 - Clause 11.2.8). The emergency shutdown button can be wired to the PLC input or to a master shutdown relay as outlined in the SRS.

10-6. Some of the key advantages are:

- Ease of programming and configuration
- Enhanced diagnostics and trouble shooting capabilities
- Cost
- More flexible for hardware and software upgrades

10-7. $\lambda^{DU} = 1500 + 13 + 125 + 75 + 125 + 75 + 125 + 75 = 2113$ FITS

10-8. Some of the main characteristics of Safety PLC are:

- Designed to fail in a predictable, safe way

- Have internal diagnostics to detect improper operation within itself.

- Have software that uses a number of special techniques to insure software reliability.

- Have extra security on any reading and writing values

Chapter 11: SIS Final Elements

11-1. a. matching valve capabilities to process requirements

11-2. d. all components

11-3. Proven in use would normally be chosen when:
- No assessment or certification exists for the component
- Good historical data exists for the component required

11-4. Approximately 87%

11-5. 88%

11-6. Yes

11-7. No

11-8. Ball surface damage

Chapter 13: Oil and Gas Production facilities

13-1. A design that is expected to satisfy the SIL 3 requirements for the HIPPS systems will be:

Sensors: 2 transmitters (1oo2 voting) with full assessment and certified as per IEC 61508. The certification is to state that the product can be used in SIL 3 applications if more than one transmitter is used and the HFT is > 0.

Logic Solver: Safety PLC certified for SIL 3 applications as per IEC 61508

Final Elements: 2 valves (1oo2 voting) with full assessment and certified as per IEC 61508 and partial valve stroke testing at a frequency at least ten times more often than the expected average demand interval.

13-2. A proof test interval of five years is used for the pressure transmitters. A proof test interval of ten years is used for the logic solver. A proof test interval of five years is used for the valve subsystem. Using a mission time of ten years and a startup time after process shutdowns of 24 hours, the following results were obtained with the SILver™ tool:

Safety Instrumented Function Performance Metrics for SIF Exercise 2 Results	Graph
Average Probability of Failure on Demand (PFDavg)	4.60E-04
Safety Integrity Level (PFDavg)	3
Safety Integrity Level (Architectural Constraints IEC 61508)	3
Risk Reduction Factor	2173
MTTFS (years)	21.83

Sensor Part Information		Graph	
Sensor Group(s)		Edit	
(1) Pressure Transmitter		Details	Graph
PFDavg Sensor Part		1.11E-06	
MTTFS Sensor Part (years)		1.90E+05	
Maximum SIL allowed (Architectural Constraints IEC 61508)		4	

Logic Solver Part Information		Graph
Logic Solver		Edit
(1) SIL 3 PLC		Details
PFDavg Logic Solver Part		3.78E-04
MTTFS Logic Solver Part (years)		194.32
Maximum SIL allowed (Architectural Constraints IEC 61508)		3

Final Element Part Information		Graph	
Final Element Group(s)		Edit	
(1) Valves		Details	Graph
PFDavg Final Element Part		8.15E-05	
MTTFS Final Element Part (years)		24.59	
Maximum SIL allowed (Architectural Constraints IEC 61508)		3	

Chapter 14: Chemical Industry

14-1. Based on the data provided for this exercise, the PFD_{avg} and other metrics pertaining to the final element subsystem is as per the table below:

PFDavg Final Element Part:	3.44E-03
HFT Final Element Part:	1
MTTFS Final Element Part (years):	7
Maximum SIL allowed (Architectural Constraints IEC 61511):	2

Chapter 15: Combined BPCS/SIS Designs

15-1. The PFD for the solenoid common cause is calculated with PFD = 1- exp(- Beta × LambdaD × TI) = 0.000008. The Probability of Solenoid A and Solenoid B failure = 0.0008 × 0.0008 = 0.00000064.

The PFD for the Solenoid system is approximately = 0.000008 + 0.00000064= 0.00000864.

15-2. To satisfy the SIL 3 requirements, three main items have to be satisfied i.e.:

- PFDavg calculations have to be satisfied
- The architectural constraints requirements have to be met
- The design and development process of the pressure transmitter subsystem needs to follow the requirements of SIL 3

15-3. LambdaS = %safe × Lambda = 0.66 × 0.06 = 0.0396

LambdaD = (1-%safe) × Lambda = (1 - 0.66) × 0.06 = 0.0204.
LambdaDD = LambdaD × C^D = 0.0204 × 0.6 = 0.01224.
LambdaDU = LambdaD × (1 - C^D) = 0.0204 × (1 - 0.6) = 0.000816

Using the approximation, PFD = LambdaDU × TI = 0.000816 × 0.5 = 0.00408

Note: LambdaDD does add to the PFD because it was assumed that those failures would automatically cause a safe shutdown.

Answer: a

15-4. c. One half of the shortest time spans should be used as a maximum. In this case answer d may also be used but it is not the maximum.

15-5. d. The hourly failure rate is 0.15/8760 = 0.000017.

Index

1oo1 Architecture 267, 318
1oo1D Architecture 333
1oo2 Architecture 267, 324
1oo2D Architecture 349
2oo2 Architecture 269, 329
2oo2D Architecture 344
2oo3 Architecture 337
3051S 136

absorbing 278
actuators 157, 159
analysis phase 8
AND gates 69, 211
annunciation 86, 129, 135, 166
ANSI/ISA-84.01-1996 2
applicability 243
approximations 49
    techniques 69
automatic risk reduction systems 19
auxiliary 99
availability 51
average probability of failure on demand 87

ball valve 162
Basic Process Control System (BPCS) 1
bathtub 33
    curve 31
Bayes' Rule 254
Bayesian approach 243
binding 161
BPCS versus SIS 20
    analysis 236
    failure mode comparison 21

censored data 37
Chi-Squared function 39
    probability distribution 38
classification 229
clean service 166
close to trip 165
complementary 246
conditional probability 250
confidence factor 38
constant failure rate 47
continuous mode 96
cut set method 259

demand 54
detected 86
diagnostic alarm 131
diagnostics 166, 305
double offset 163
down-hole safety valve (DSV) 192

emergency shutdown 2
end of life 31
EPA/OSHA findings 5
equipment
    classification 230
    justification 230
    selection 91
ergotic 278
event space method 257

fail safe 2, 83, 85, 159
fail-danger 83, 85
failure 27
    databases 118

modes 83
failure modes and effects analysis (FMEA) 303
failure modes effects and diagnostics analysis
   (FMEDA) 92, 121, 148, 306
failure rate 30
   comparison 122
   estimating 34
   time-dependent 30
fault
   avoidance 94
   control 94
   tree 65, 111, 199, 257
   tree symbols 67
final elements 84, 157, 175
flame detectors 142
floating ball valve 163
flow measurement 140
flow-line valve (FLV) 193
FMEA
   format 304
   procedure 303
FMEDA
   assessment 92
   limitations 312
frequency of occurrence 245
full stroke testing 170
functional failures (see systematic failures) 28
functional safety 1
   equipment 91
   standards 2, 29

gas
   blowby 225
   detectors 141
gate
   solution method 264
   valve 164
globe valve 164

high demand mode 97, 102
high performance butterfly valve 163
hydraulic 159
   controls 158
   pressure 193

IEC 61508 3, 85, 93, 107, 136, 148, 183
imperfect testing 56
independent events 248
infant mortality 31
input circuit 150
input module 150
instrument selection 129
intended function 61
intersection 246

intrinsically safe 146

leakage 161
level 138
   switches 138
   transmitters 138
logic 195
   solver 145, 175
low demand mode 97, 102
lower master trip valve (LMV) 193

main processor 150
manual proof test effectiveness 217
Markov model 53, 74, 112, 133, 275, 321
   solution techniques 75
mean 242
Mean Time Between Failures (MTBF) 51
mean time to fail dangerously 87
Mean Time To Failure (MTTF) 46
Mean Time To Restore (MTTR) 50
Mean Time To Trip Spuriously (MTTFS) 11
median 242
memory-less 275
model building 61
multiple failure state components 290
mutually exclusive sets 247

no effect 86
non fail safe 159
non-repairable component 284
normal diagnostics 121

oil and gas production facilities 189
open to trip 165
operation phase 11
OR gates 69, 211
output circuit 150

parallel network 63
partial stroke testing (PST) 185
partial valve stroke testing 167
PFD 87
PFDavg 87, 112, 218, 266
PFS 87
physical failures (see random failures) 28
pneumatic 159
   controls 158
   logic 145
pollution control equipment 5
pressure 135
pre-startup safety review (PSSR) 11
prior use assessment 93
probabilistic calculation tools 112
probabilistic modeling 150, 165

probability   245
   dangerous failure per hour   102
   failing safely   87
   failure on demand (PFD)   87, 199
   success   43
   summation   249
process equipment reliability database   122
process hazards analysis (PHA)   4, 216
process safety review   29
programmable logic controllers (PLCs)   2, 146
PST credit   168

random
   failures   28, 117
   variables   239
reactors   215
realization phase   9
redundancy   94
relays   145
reliability   43, 45, 99
   block diagrams   62, 68, 257
   engineering   43
   function   44
remote actuated valve   158
repairable component   282
resilient butterfly valve   163
restoration   56
risk reduction factor   87

safety
   functional   1
safety instrumented function   23, 27
   equipment   23
   personnel   24
safety instrumented systems   1–2, 19
safety integrity level (SIL)   173
safety lifecycle   1, 5
   adoption   13
   benefits   12
safety manual   94
safety metric calculation   99
safety PLC   147–148, 180, 183, 197, 204
safety relays   145
safety requirements specification (SRS)   89
safety transmitter   180
safety verification   150
satisfactory performance   44
sensors   129, 175, 195
separation vessels   223
separators   191
series system   62
severe service   166
shutdowns   191
shutoff   165

SIF
   identification   99
   modeling   86
   PFDavg calculation   196
   testing techniques   96
   verification   33
   verification analysis   67, 209
SIL 1 architecture   176
SIL 2 architecture   180
SIL 3 Architecture   185
SIL
   capability   94
   verification   227
SIL2
   architecture   185
   requirements   180
single shutdown valve   180
SIS
   engineering requirements   22
   operating conditions   22
software
   criticality analysis   148
   fault injection   148
solid state   2, 146
specific hazardous event   23
standard deviation   243
statistical analysis   239–240
statistical significance   243
steady state
   availability   52
   probability   280
stems   161
stress   33
susceptibility   33–34
system architectures   316
systematic failures   28, 117

temperature   137
tight shutoff   165
time dependent analytical solutions   284
time to failure   30
trip amplifiers   145
triple offset butterfly valve   163
trunnion mount ball valve   162

unavailability   52
undesirable event   65
undetected   86
union   246
unreliability   46, 63
upper master valve (UMV)   193
useful life   31

valve failure modes   160

valves 157, 196
    ball 157
    butterfly 157
    gate 157
    globe 157
    offset butterfly 157
Venn diagram 245
verification reports 112

warnings 5
wearout region 31
well controls 191
well designed 118, 158–159
wing valve 193